RESI
AUDIO/VIDEO SYSTEMS
ENDORSEMENT

Copyright © 2009 by Educational Technologies Group

This book was set in Futura MD BT, Times New Roman, and Arial by Cathy J. Boulay, ETG.

Written by   Max Main

Graphics by   Michael R. Hall
Tristan J. Weetch

ISBN-13: 978-1-58122-103-9
ISBN-10: 1-58122-103-7

1 2 3 4 5 6 7 8 9 10   13 12 11 10 09

PREFACE

PURPOSE

Marcraft International has been producing certification courseware for different facets of the electronic, computer architecture, computer repair, and IT industry since 1988. This book has been designed to prepare students for the ETA RESI Audio and Video Certification exam, and provide them with the necessary tools for grasping the complex technologies of home automation.

The science of automated home technology has captured the imagination of consumers, standards groups, equipment suppliers, and home designers around the world. In only a few years, the home automation industry has grown rapidly from a hobbyist pastime to a mature technology that promises to provide increased efficiency in such areas as home energy management, heating and air conditioning management, home office networking, telecommunications, entertainment systems, irrigation systems, security systems, residential access systems, and more.

The home office has become a valuable and necessary asset for many homeowners. The introduction of wireless networking products and the "no new wires" network design standards have opened the door for new, low-cost design alternatives for the home computer user. Existing phone lines, or power line wiring, can now serve as the home network media supported by a wide array of industry standardized "plug and play" devices.

New Telecommunications Industry Association (TIA) standards and Federal Communications Commission (FCC) rulings issued in recent years have recognized the growing need for residential, as well as commercial, wiring standardization. The emergence of TIA-endorsed structured wiring designs for the whole home has rapidly become the main platform for integrating all of the home control and automation features into a cohesive home management system. The FCC has established a minimum grade for residential telecommunications cable installed in new homes. Until a few years ago, each residential automation or control technology required its own type of wiring and unique topology. Now, a single structured wiring technology standard has established a goal for supporting all the major new home automation technologies.

With the rapid growth of the Internet, faster desktop PCs, and the emergence of streaming audio and video, MP3, online game playing, e-mail, and improved residential broadband Internet access methodologies, the home office user can now utilize many of the advanced networking features previously available primarily for the corporate user.

It is now the task of the home technology integrators who are using this book to master the art of designing, installing, and programming the various home controllers, gateways, routers, home theater systems, and telecommunications devices that will rapidly become the core feature in the "smart home" of tomorrow.

RESI AUDIO AND VIDEO CERTIFICATION

The Electronics Technicians Association, International (ETA®) is an organization that establishes certification criteria for service technicians in the computer industry. With the infusion of IT into consumer electronic products, digital systems have become more popular and more powerful than ever before.

Please visit *www.etainternational.org* to view the ETA exam competencies and practice exams for audio and video systems.

Examinees have two ways to sit for a certification exam with ETA International. They may take an exam using either a paper or online format under the supervision of an ETA approved exam proctor. Exam results are returned to examinees within ten business days when paper exams are received at ETA's home office. Immediate results are displayed when the examinee completes the online exam.

ETA International currently has 850 ETA Certification exam proctors across the US and worldwide. To take an exam, locate an exam administrator (CA) and exam fees, please visit *www.etainternational.org* or call 1-800-288-3824 for more information. If your educational facility is interested in proctoring ETA exam, please call 1-800-288-3824 or e-mail us at *eta@eta-i.org*.

Key Features

The pedagogical features of this book were carefully developed to provide readers with key content information, as well as review and testing opportunities. A complete RESI Audio and Video objectives map and various Test Tip information boxes are included to help students key in on RESI specific materials.

RESI Audio and Video Exam Coverage

The locations of RESI Audio and Video specific materials are identified with a Test Tip marker in the margin that helps students focus on key content that they will be expected to know for the RESI Audio and Video Certification exam. Appendix A provides a comprehensive listing of the RESI Audio and Video objectives.

Pedagogical Features

A wealth of figures, diagrams, and screen dumps are included to provide constant visual reinforcement of the concepts being discussed.

The book begins with a list of learning objectives that establishes a foundation and systematic preview of the book. In addition, each book concludes with a book summary and a key-point review of its material.

Key terms are presented in bold type throughout the text. A comprehensive glossary of terms provides quick, easy access to key term definitions that appear in each book.

In general, it is not necessary to move through this text in the same order that it is presented. Also, it is not necessary to teach any specific portion of the material to its extreme. Instead, the material can be adjusted to fit the length of your course. As a matter of fact, practicing IT professionals can use the material in the CD test banks to identify the areas where they need to brush up, and then use the text book to study that material directly.

Teacher Support

An instructor's guide accompanies the course. Answers for the end-of-book quiz questions is included along with a reference point in the book where a particular item is covered. Sample schedules are included as guidelines for possible course implementations. Answers to all lab review questions and fill-in-the-blank steps are provided so that there is an indication of what the expected outcomes should be.

An electronic copy of the textbook is included on the Instructor's Guide CD-ROM disc.

Test Taking Tips

The RESI Audio and Video Certification exam is an objective-based timed test. It covers the objectives listed in Appendix A in a multiple-choice format. There are two general methods of preparing for the test. If you are an experienced technician using this material to obtain certification, use the testing features at the end of the book and on the accompanying CD to test each area of knowledge. Track your weak areas and spend the most time concentrating on them.

After completing the study materials use the various testing functions available on the CD to practice taking the test. Use the Study and Exam modes to test yourself by chapter, or on a mixture of questions from all areas of the text. Practice until you are very certain that you are ready. The CD will allow you to immediately refer to the area of the text that covers material that you might miss.

- Answer the questions you know first. You can always go back and work on previous questions.

- Don't leave any questions unanswered. They will be counted as incorrect.

- There are no trick questions. The most correct answer is in there somewhere.

- Be aware of questions that have more than one correct answer. Questions that have multiple correct answers can be identified by the special formatting applied to the letters of their possible answers. They are enclosed in a square box. When you encounter these questions, make sure to mark every answer that applies.

- Get plenty of hands-on practice before the test, using the time limit set for the test.

- Make certain to prepare for each test category listed above. The key is not to memorize, but to understand the topics.

- Take your watch. The RESI Audio and Video Certification exam is a timed test. You will need to keep an eye on the time to make sure that you are getting to the items that you are most sure of.

- Get plenty of rest before taking the test.

All ETA exams are closed-book exams, which means that you will not be permitted to take any material into the testing area. However, you will be provided with a blank sheet of paper and a pen or pencil. When you complete a ETA certification exam, the test center software informs you whether you have passed or failed the exam.

Table of Contents

AUDIO AND VIDEO SYSTEMS

LAB PROCEDURES – AUDIO AND VIDEO FUNDAMENTALS

LAB PROCEDURE 1 – HOME AUDIO/VIDEO SYSTEM DESIGN

LAB PROCEDURE 2 – INSTALLING A MULTIZONE AUDIO SYSTEM

LAB PROCEDURE 3 – ADJUSTING LCD TV MONITORS AND SURROUND SOUND AUDIO

AUDIO AND VIDEO SYSTEMS

U pon completion of this book and its related lab procedures, you should be able to accomplish the following tasks.

- Implement, maintain and troubleshoot multiroom audio systems.
  - Control devices
  - Differentiate and define single source, multi-source, and local source.
  - Amplification
  - Speaker types
  - Speaker specifications

- Install, configure and maintain a residential home theater system.
  - Audio Components
  - Video components

- Assess, install and configure content management systems and describe their applications in a residential environment.
  - Describe typical applications and physical connections of sources
  - Summarize types of media storage, methods to transfer and backup data.
  - Other connection considerations

- Implement, maintain and troubleshoot multiroom video systems.
  - Define signal types and their applications
  - Identify cable types and their applications
  - Termination (e.g. RCA, BNC, and F)
  - Satellite

Audio and Video Systems

INTRODUCTION

This book covers the basic features, products, and design concepts for home audio and video systems.

The home entertainment industry has expanded rapidly in recent years with the introduction of several new technologies and products. Home theater designs now compare favorably with commercial movie theater sound quality and screen displays. Surround sound and High-Definition TV (HDTV) bring a new dimension to the residential audio and video system capability.

This book is structured to identify and describe many of these innovative and emerging standards and technology associated with designing and installing residential audio and video components. The key objectives outlined and covered in this book will assist the home technology integrator to select the best options when planning and installing home audio and video systems, satellite antennas, and home entertainment systems.

AUDIO/VIDEO BASICS

One of the most significant consumer electronics subsystems in a typical residential setting is the **audio/video (A/V) system**. Some type of electronic audio and video systems can be found in nearly every home. These systems may be very simple—made up of a stand alone radio receiver and a stand alone television set, or they may be as complex as a special home theater room with distributed whole-house audio and video systems. In either case, the system can be simplified into two major concerns—the system's components and its connections.

The components that make up an A/V system can be divided into three basic types of equipment, as illustrated in Figure 1-1:

- Sources
- Processors
- Content Management Devices
- Output Devices

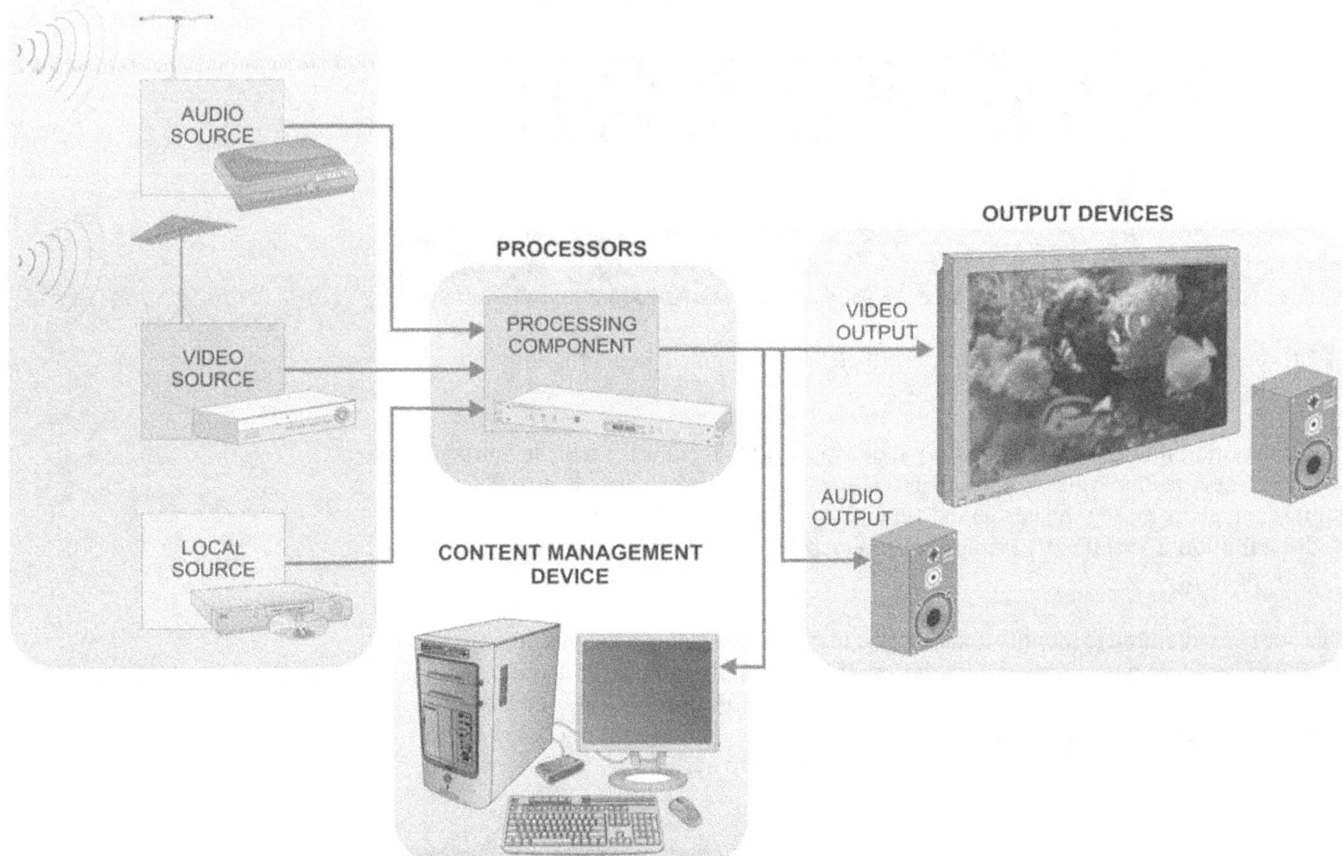

**Figure 1-1: A/V
System Components**

Sources

A/V sources

receivers

players

A/V sources are devices in the A/V system where the audio or video signal originates. These devices include two major types of devices—**receivers** and **players**. Receivers are devices that obtain audio or video information from somewhere else—such as an outside service, as depicted in Figure 1-2. Players are devices that obtain audio or video information from a local media, such as tape, optical disc, vinyl records or electronic memory devices, as shown in Figure 1-3.

Typical A/V receivers include:

- AM/FM Radio receivers

- Off-Air Television receivers

- Cable Television receivers

- Satellite Television receivers

- Satellite Radio receivers

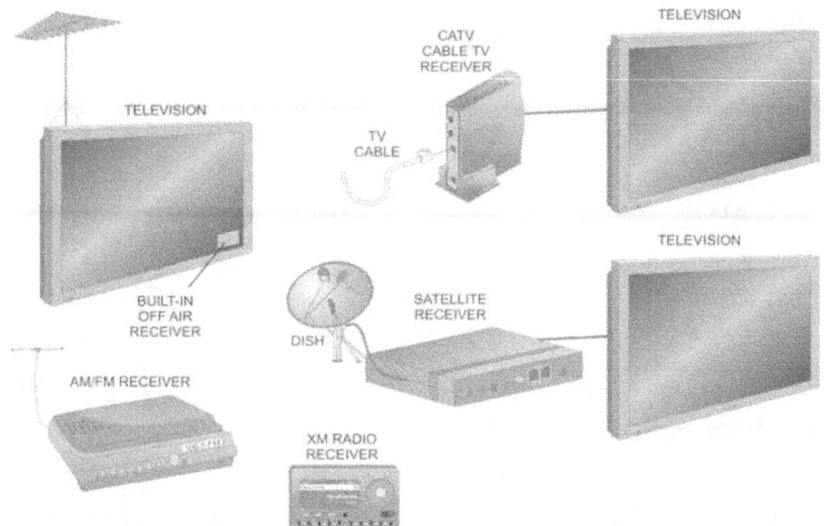

Figure 1-2:
A/V Receivers

Typical A/V players include:

- CD/DVD disc players

- Tape players

- Record players (turntables)

- MPEG players

- Game Consoles

A/V sources may also be classified by the type of information they provide. These classes are audio-only sources—that only provide sound—and audio with video sources that provide both sound and video information.

Examples of audio-only sources include:

- CD player

- Tape deck

- Turntable

- MPEG player (Apple iPod, etc.)

- Satellite radio (Sirius, XM)

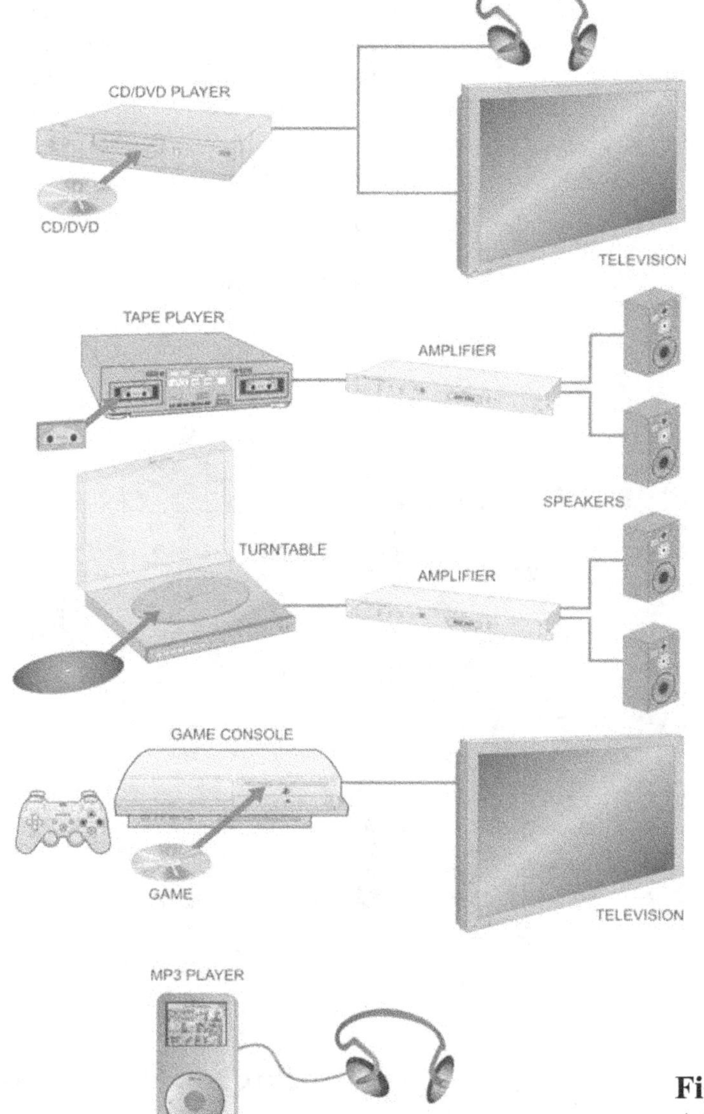

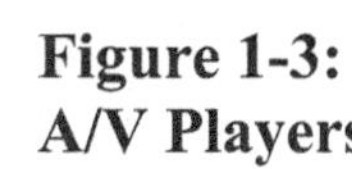

Figure 1-3:
A/V Players

Examples of audio-with-video sources include:

- VCR
- DVD player
- Satellite box
- Cable box

- Digital Video Recorder
- DVD recorder
- XBOX, Playstation

Signal Processors

Signal processors are devices or circuits that change the format of the audio or video signal, as illustrated in Figure 1-4. They may increase the strength of the signal for transmission or some particular application. In addition, signal processors may extract information from an encoded signal, select between different signals, adjust signals for specific purposes, or change the level or format of the signal to make it more suitable for transportation from one pace to another. Devices in this category include:

- Amplifiers
- Decoders
- Modulators

- Multiplexers
- Equalizers

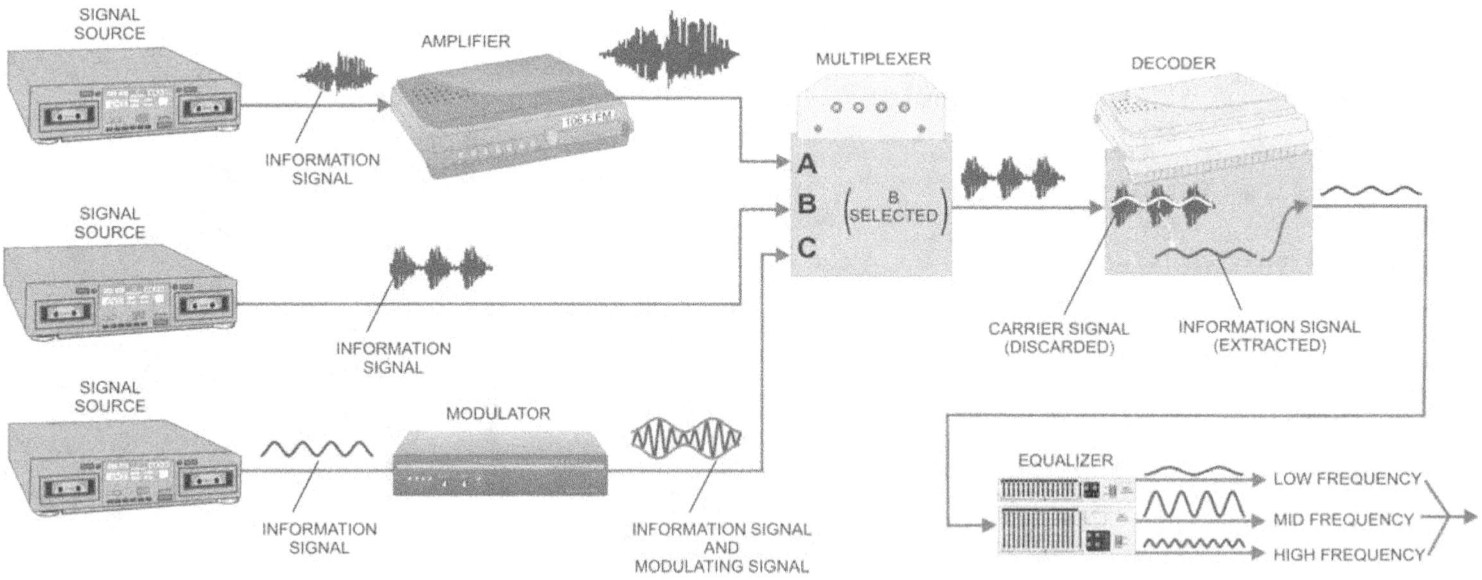

Figure 1-4: Signal Processors

Content Management Devices

Many modern home entertainment configurations include a **Content Management System** that offers centralized control of the A/V system as well as storage and management of the video and audio information. This requires an intelligent management system that is often based on a personal computer and special software applications designed to handle digitized documents, pictures, video, movie and audio files. These systems normally include both local and remote controller options that enable users to interact with the system in the most convenient way. Typical A/V content management systems, like those depicted in Figure 1-5, include:

- Digital Video Recorders
- Content Management Devices
- Controllers

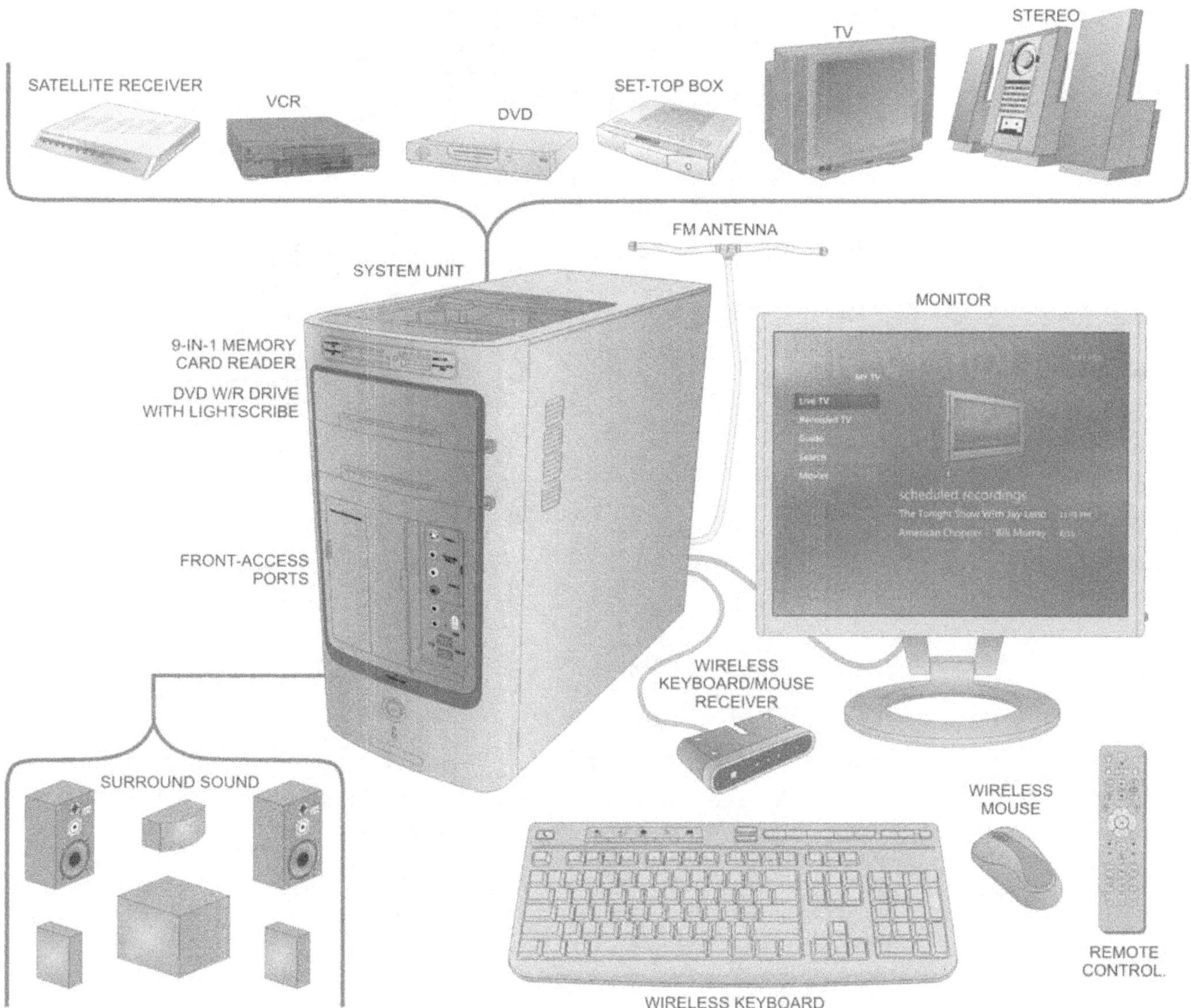

Figure 1-5: Content Management Systems

speakers

video displays

Output Devices

A/V output devices are those components that actually deliver the audio or video information to the user, as illustrated in Figure 1-6. These devices come in two basic forms—**speakers** that deliver *audio information* to a listener and **video displays** that deliver *video information* to a viewer.

- Speakers
- Video Displays

**Figure 1-6:
Speakers and Displays**

Hybrid Devices

hybrid devices

Some devices serve more than one function—these are referred to as **hybrid devices**. A television with integrated speakers could handle the receiver, signal processing and output functions for both sound and video. You could also have an AM/FM receiver that includes a tuner, a built in amplifier and equalizer that provides the receiver and two signal processing functions in a single component. In this case external speakers would handle the output function, as shown in Figure 1-7.

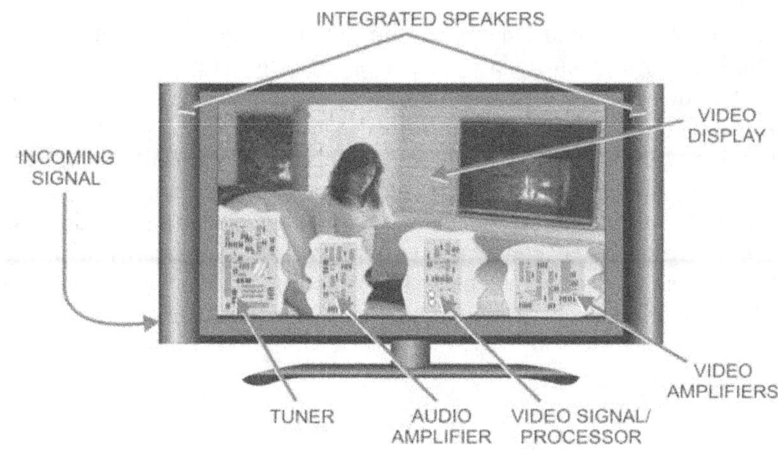

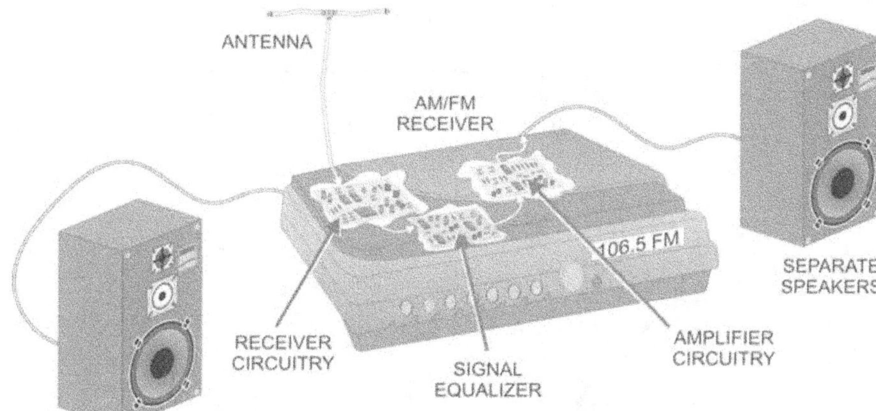

Figure 1-7:
Hybrid A/V Devices

Signal Paths and Connection Types

Most residential A/V equipment is modular in design. This allows the A/V consumer to mix and match different types of A/V components in their home to meet their entertainment and budget needs. What this means for the professional A/V installer is that there may be many different options for interconnecting the customer's A/V components.

When tying together different devices, it is important for the installer to plan out how the **signal** will get from where it originates to where it needs to go. Understanding the signal path is not only crucial for optimal performance, but also in avoiding the need for troubleshooting.

The computer industry has developed and standardized a number of different connectivity options. In some cases there are even competing standards for the same functions (high-speed USB and FireWire serial connections). Likewise, the consumer electronics industry has developed its own connectivity standards and options. You will need to understand these standards and options to create the best entertainment experiences for your customers.

For instance, you may have a situation where a signal-processing component offers different connectivity options for a given function (such as video in). However, one option offers superior characteristics over the other. If you have two video source components that must run through the processing component, you will need to decide which source should use the connectivity option with the superior qualities. At this point, you will need to access the components involved and how the customer plans to use the system. Figure 1-8 shows a connectivity scenario where multiple video devices are connected to a video display device. The video display offers different connection options with differing quality. In this scenario, the simple DVD player has been connected to the composite input option while the AM/FM receiver/DVD player has been given the higher quality S-video connection.

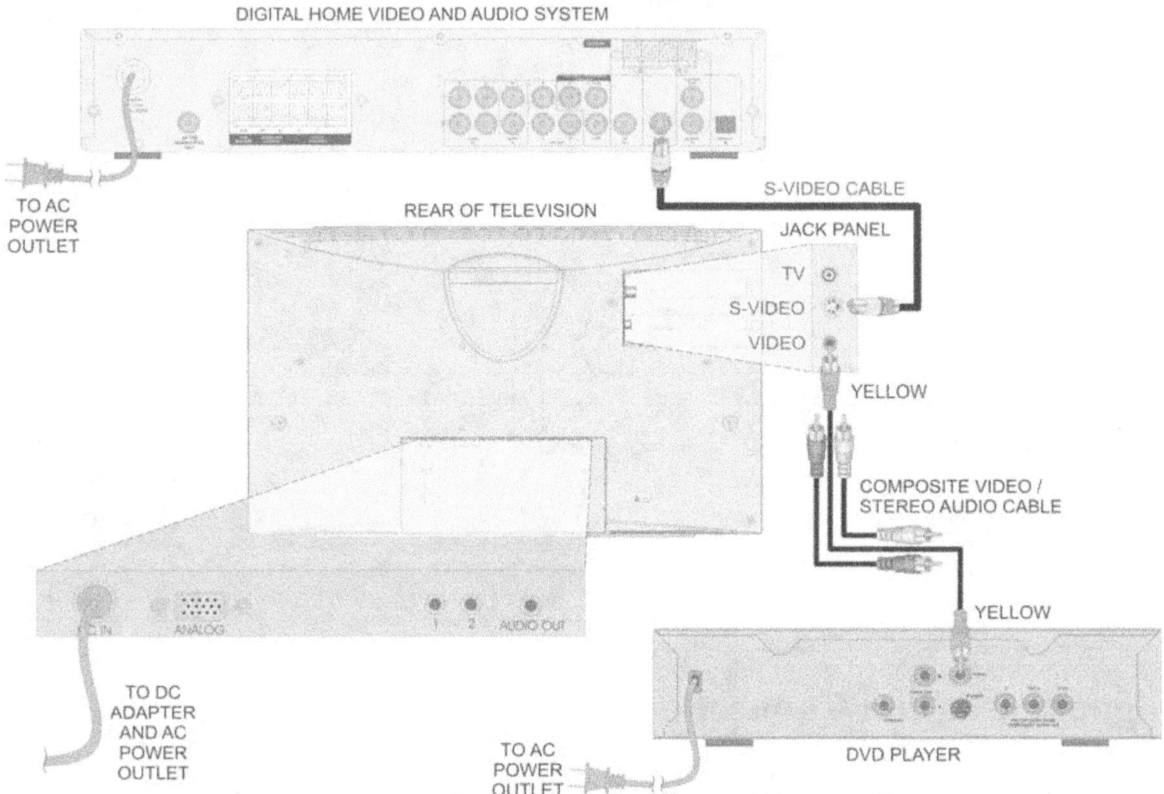

**Figure 1-8:
Video Display
Device**

AUDIO AND VIDEO COMPONENTS

This section covers the most common types of audio and video components used in whole-home and dedicated designs. The components included are:

- A/V receivers
- Amplifiers
- Equalizers
- Modulators
- Speakers
- TV display configurations

- Touchscreens
- DVD players and recorders
- Laser disc players
- Personal Video Recorders (PVRs)
- Keypads and remote controllers
- Satellite receivers and antennas

Receivers

A **receiver** is one of the main components of an audio/video system. It is used to process audio, video, satellite, and AM/FM broadcast signals received from external sources and to output amplified or decoded signals to television video input jacks, speakers, or recording devices.

These devices are routinely equipped with various rotary and pushbutton controls for volume, tone, and other sound enhancements and adjustments. Most receivers used in home audio/video systems combine the functions of a radio tuner, preamplifier, and amplifier in a single chassis. Receiver systems are also available in which the preamplifier, tuner, and amplifier are separate units. Although the separate units require more shelf space, they usually provide more features and options that can affect or enhance sound quality. A block diagram of a receiver with the three embedded component systems is shown in Figure 1-9.

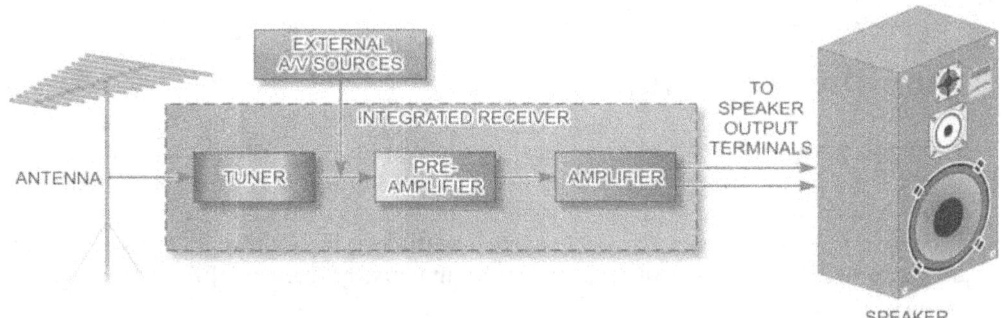

Figure 1-9:
Audio/Video Receiver

Most receivers are sold under various names with varying capabilities and configurations. The common types of receivers are described as follows:

- Audio/Video (A/V) receivers have an integrated AM/FM tuner and process from 4 to 6 channels with amplifiers capable of driving multiple speaker channels. They have built-in surround sound decoders that separate the sound signal from a video source (like a DVD player) into several channels, then route it to several speakers. There are numerous surround sound formats (covered later). The most common types of surround sound decoders are Dolby Pro-Logic, which processes four channels of sound, and Dolby Digital, which processes six channels known as the 5.1 format. A/V receivers are also available in Dolby Pro-Logic/Digital Ready and Dolby Digital 5.1/DTS (Dolby Theater System) modes. A/V receivers usually have both audio and video input jacks for DVD, S-video, component, and composite video.

- Stereo receivers are less sophisticated than A/V receivers. They also have an AM/FM integrated tuner for receiving commercial broadcast stations. Stereo receivers split the signal from an audio source, such as a CD player, into two channels, left (L) and right ®), then route it to two speakers. Many stereo receivers have outputs for an extra set of speakers, but are limited to two channels of sound. Stereo receivers are the least expensive type of receivers. They lack the capability for full surround sound channel processing found in later model A/V receivers.

- Satellite receivers are used to receive and process high-frequency digitally encoded signals from a satellite dish antenna. They demodulate and decode digital television programs received from Direct Broadcast Satellite (DBS) service.

Amplifier Output Power Specifications

All power amplifiers are designed to deliver an output signal rated in watts. The power is usually rated for various load impedances. The most common is 8 ohms, with 4 ohms and 2 ohms also sometimes listed.

These impedances are designed to match the load impedance of speakers connected to the output of the amplifier. Impedance ratings for speakers are not the same as resistance although both are measured in ohms. The value of impedance varies with the frequency of the signal out of the amplifier. The impedance of a speaker is its opposition to alternating current (AC). The lower the impedance, the higher the current, therefore the speaker impedance will vary above and below the nominal impedance when audio is amplified and passed to the speaker for reproduction. Ohms are the unit of measurement for what is normally listed as the "nominal impedance."

Amplifier Bridging

Bridging an amplifier is a technique used to configure a two-channel stereo amplifier to drive a single load (speaker) with more power than the sum of the two original channels. An amplifier running in bridged mode has a single output channel to which a load (speaker) can be connected. It is no longer a two-channel (stereo) amp as far as the input signals and loads are concerned. If you have two speakers and want to use bridged amplifiers, you will need *two* stereo amplifiers.

With bridged stereo amplifiers a single input signal is applied to the amplifier. Inside the amplifier the signal is split into two signals. One is identical to the original, and the second is *inverted.* The original signal is sent to one channel of the amplifier, and the inverted signal is applied to the second channel. The output produces two channels, which are *identical* except that one channel is the *inverse* of the other.

A bridged amplifier may be used to increase the power available to a single speaker because the output voltage is effectively doubled, which theoretically increases the power output by a factor of 4.

As shown in Figure 1-10, the speaker is connected between the left channel *positive* output terminal and the right channel *negative* output terminal.

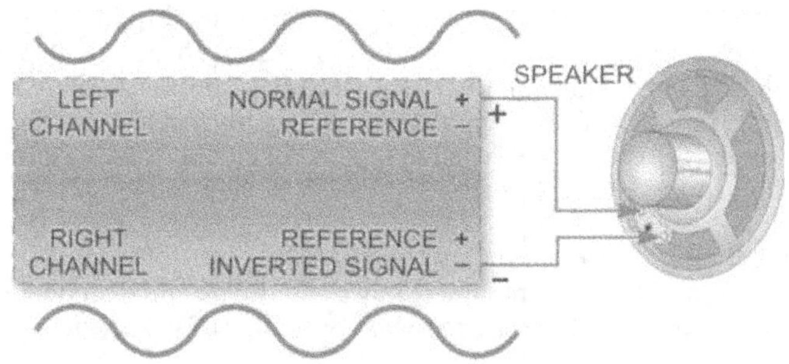

**Figure 1-10:
Bridged Amplifier**

In other words, one channel "pulls" one way while the second channel "pulls" in the opposite direction. This delivers considerably more power to a single load than is the case with a single stereo amplifier. Amplifiers running in bridged mode are typically limited to speakers with impedance ratings of no less than 8 ohms.

Know that a bridged amplifier can be used to increase the power available to a single speaker because the output voltage is effectively doubled, which theoretically increases the power output by a factor of 4.

Sound System Equalizers

Equalization is a sound processing function that selectively boosts or decreases bands of audio frequencies to match room acoustics to the sound system.

The most common equalization systems used in home audio systems are the tone controls. They provide a quick and easy way to adjust the sound and partially compensate for the room acoustics. These controls are labeled "bass" and "treble." Setting the equalizer is ultimately a matter of individual user preference.

Graphic equalizers, as shown in Figure 1-11, are a step up from tone controls in terms of capability and control. A graphic equalizer is simply a set of filters, each with a fixed center frequency that cannot be changed.

The only control you have is the amount of boost or cut in each frequency band. This boost or cut is most often controlled with sliders. This interface is fairly intuitive because the frequency response of the equalizer resembles the positions of the sliders themselves. The sliders are a graphic representation of the frequency response, hence the name "graphic" equalizer.

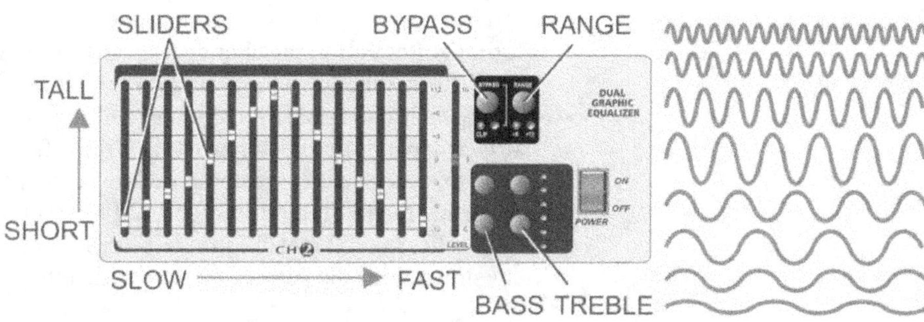

Figure 1-11: Graphic Equalizer

Modulators

Modulators are used in video distribution systems. They are available in single- or multiple-channel configurations. A multi-channel modulator is shown in Figure 1-12. They provide internal television channel selection for the internal broadcasting of local video sources such as a VCR, a laserdisc, a DVD, or CCTV cameras. It is possible to "combine" these new TV channels with existing cable or off-air channels and easily distribute all signals throughout the home on a single coaxial cable.

Know the purpose of a modulator used in a whole-home video distribution system.

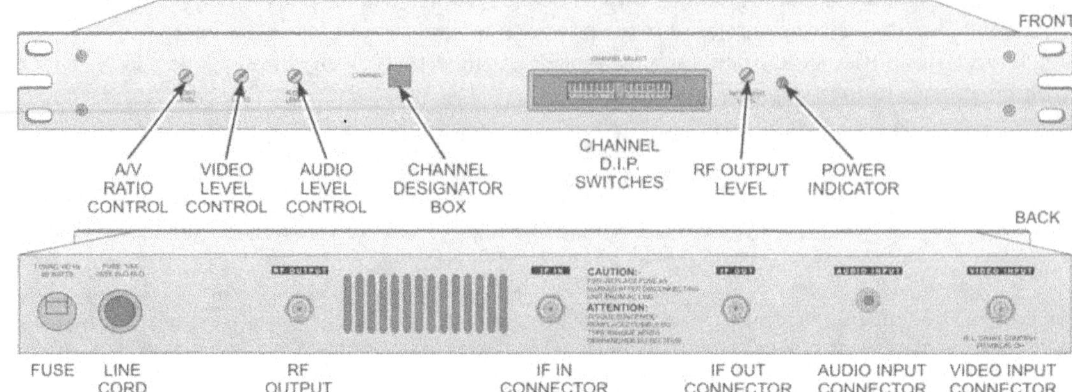

**Figure 1-12:
Multi-Channel
Video Modulator**

AUDIO SPEAKER DESIGN SOLUTIONS

Speakers and their enclosures are designed based upon the fundamental laws of acoustics. Acoustics can be thought of as the scientific study of sound, including the characteristic way in which sound carries or can be heard within an enclosed space, such as a room or a speaker enclosure. In addition to involving the placement of various speakers within an enclosure or within a room, important parameters also include sound reflection, cancellation, and balance.

Sound reflection is the change in direction of a sound wave front at an interface between two different media. This results in the wave front returning towards or into the medium from which it originated. Coherent reflection of a longitudinal sound wave striking a flat surface depends on the dimensions of the reflective surface. More notable reflections occur when this surface is large compared to the wavelength of the sound. Due to the rather wide frequency range of audible sound (20 to 20,000 Hz), these wavelengths can vary between 0.68 inches to 56.2 feet (17 mm to 17 m). Their reflections will vary according to the texture and structure of the surface against which they strike.

This means that some of the sound energy will be absorbed by porous materials, while hard and roughened materials will tend to reflect it in various directions. In this case, the sound energy will be scattered, and not coherently reflected. The auditory character of an architectural space is determined by the nature of these reflections, and this is where the field of acoustics plays a critical role. In attempting to implement exterior noise mitigation, overly reflective surfaces must be avoided. However, sound reflections can be used within speaker enclosures to directionally guide and control the desired sound.

Incoming sound waves can be electronically altered in such as way as to minimize or eliminate them altogether. Known as **sound cancellation**, the sound wave is first detected (by a microphone), and then analyzed by a microprocessor. By placing a speaker in the path of the sound wave, and broadcasting a mirror image (exact opposite) frequency, the offending wave can be significantly flattened or completely cancelled out!. This technique has proven effective in controlled environments featuring few frequencies to mirror, such as with noise-canceling headphones. However, large open areas require an expensive amount of processing to analyze and mask so many undesired frequencies.

Undesired sound cancellation can occur when wire connections to one speaker have been erroneously reversed. By reversing the phase of one speaker's sound wave in reference to than of another speaker, the sound waves will destructively interfere, resulting in diminished volume. This cancellation phenomenon is common to all wave events, including sound waves, electrical signals, waves in water, and light waves. Even properly phased speakers can cancel each others sound waves if they are improperly oriented with respect to each other and the room in which they are placed.

If the room features hard parallel walls, it may be necessary to artificially reduce any repetitive reflections between them, through the use of acoustical absorption panels. Although the use of artificial absorption can reduce these reflections, it also can result in acoustically dead areas in the room. This causes a lack of **sound balance**, reducing the intelligibility of acoustic speech levels and impairing the sound quality across the spectrum. The proper mounting of acoustical hardwood molding can provide control of unwanted reflections, as well as exceptional bass absorption. Combined with sound diffusion panels, small rooms with flat parallel surfaces can be converted into functional spaces featuring exceptional sound balance, with a natural, comfortable ambiance.

Speakers produce sound from the electrical audio signal developed by the amplifier from various audio entertainment source components such as radio, TV, VCRs, CDs, and DVDs. The goal of an audio system is to reproduce sounds in a manner that matches the original signal as accurately as possible. In this section, we will discuss how this is accomplished. Figure 1-13 shows the components as depicted with a typical speaker.

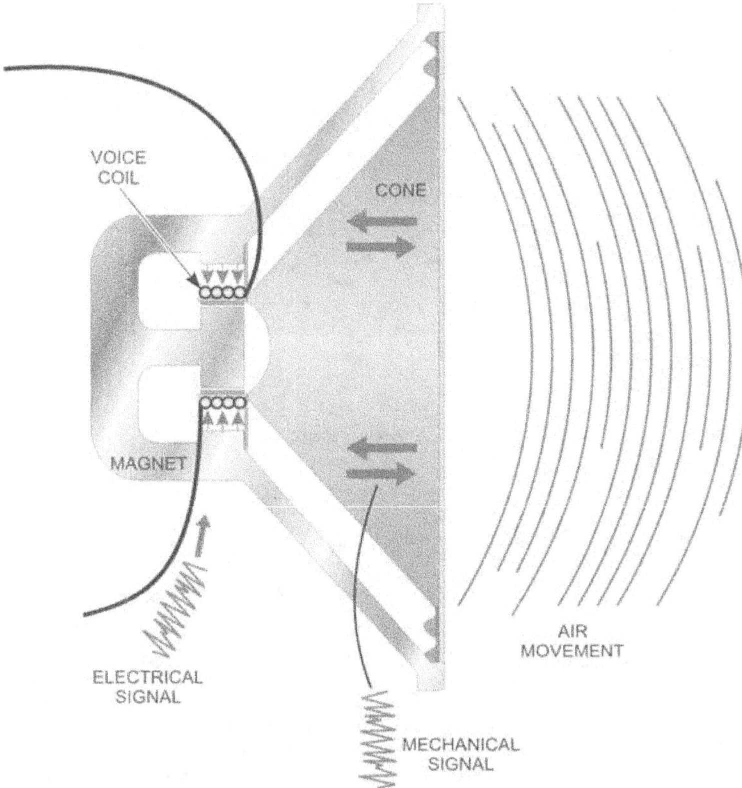

Figure 1-13: Anatomy of a Speaker

A **driver** is the part of a speaker that converts an electrical current into sound. When the coil is energized, it is repelled from or attracted to a magnet based on the polarity of the current being applied. The coil will then push or pull the driver accordingly. The driver moves the air in front of it, and this air movement is perceived by us as sound.

Speakers are available in various physical design, cost, and location options. Dynamic cone loudspeakers are the most common type of speaker used in home audio systems. Several speakers can be installed in a home theater surround sound configuration to take advantage of multiple sound tracks in DVD movies and HDTV broadcasting. High-performance speakers such as electrostatic and ribbon speakers depart from the traditional dynamic speaker design configuration but are considerably higher in cost.

Dynamic Speakers

The **dynamic speaker**, also referred to as a *cone speaker*, is a type of speaker in which the audio signal is applied to a voice coil that is part of the moving system. As indicated in Figure 1-14, the voice coil is mounted in a way that allows it to move freely within the strong field of a permanent magnet. The cone assembly is attached to the voice coil and the cone is also attached to the outer suspension ring. The speaker cone moves air back and forth when the audio signal is applied to the voice coil, which in turn produces an audible sound reproduction of the electrical audio signal.

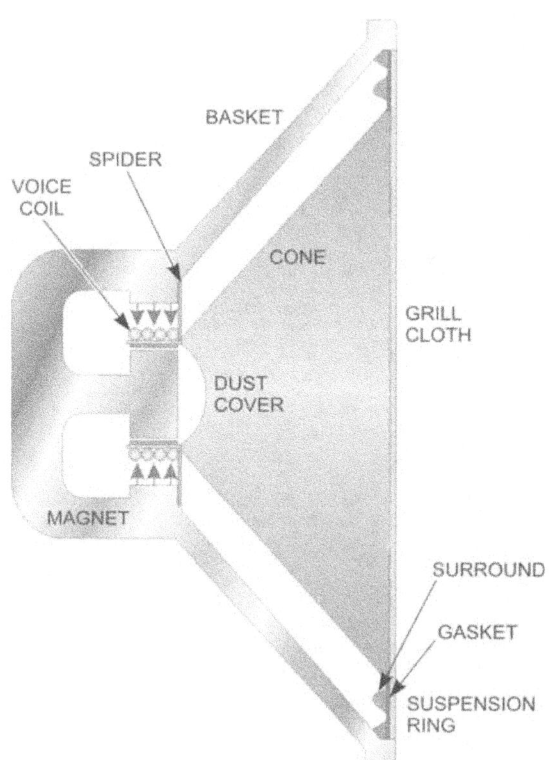

Figure 1-14:
Dynamic Speaker

Electrostatic Speakers

Electrostatic speakers are high-performance speaker systems designed to avoid the problems inherent in cone speakers. They use a graphite-coated plastic membrane suspended between two perforated metal sheets. A high voltage of several thousand volts is applied to the membrane. The input signal is also raised to a high voltage by a transformer and is applied to the perforated metal sheets. A typical electrostatic speaker is shown in Figure 1-15.

Ribbon Speakers

Ribbon speakers are another type of high-performance speakers that do not use a cone and voice coil. They use thin foils suspended between magnets or metal sheets. The input signal from the amplifier is applied to the foil ribbon. The varying electrical charge caused by the input signal forces the foil ribbon to be repelled or attracted to the magnets, thereby moving air and producing sound. It is a design that overcomes some of the deficiencies of cone speakers in the mid and upper range of the audio spectrum. Ribbon speakers like the one shown in Figure 1-16 also have a thin panel design form similar to electrostatic speakers. Both ribbon and electrostatic speakers are more expensive than cone speakers.

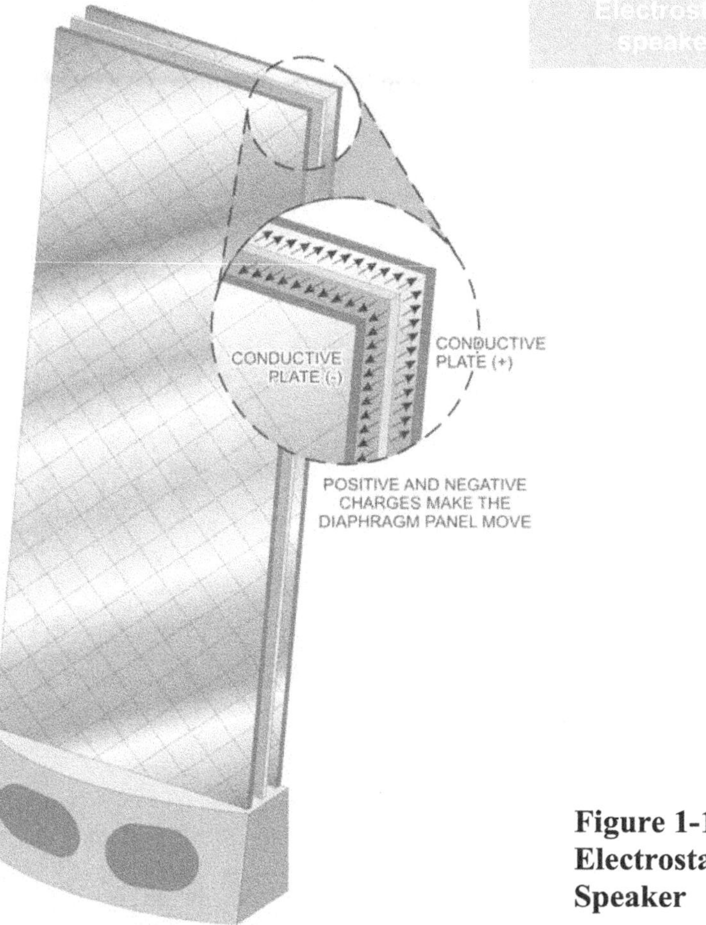

Figure 1-15: Electrostatic Speaker

Electrostatic speakers

Ribbon speakers

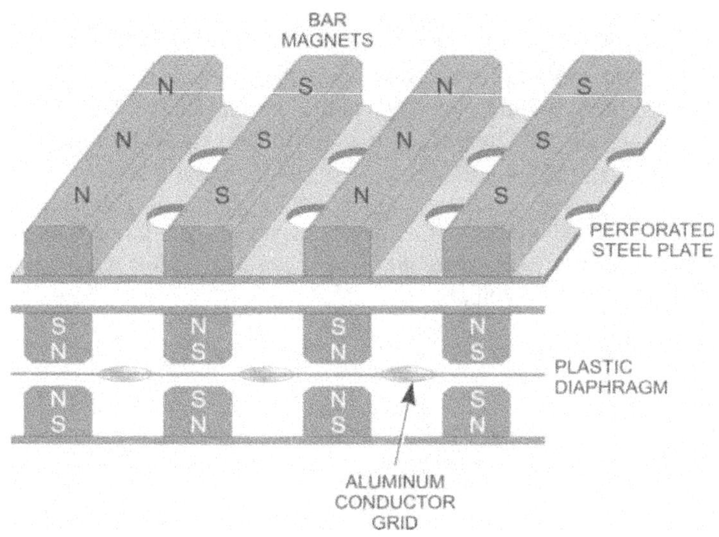

Figure 1-16: Ribbon Speaker

In-Wall and Ceiling-Mounted Speakers

In-wall speakers work like regular dynamic cone speakers, but they use the wall as a baffle. Wall speakers, as shown in Figure 1-17, can be mounted in various locations in a home multizone, multisource audio system.

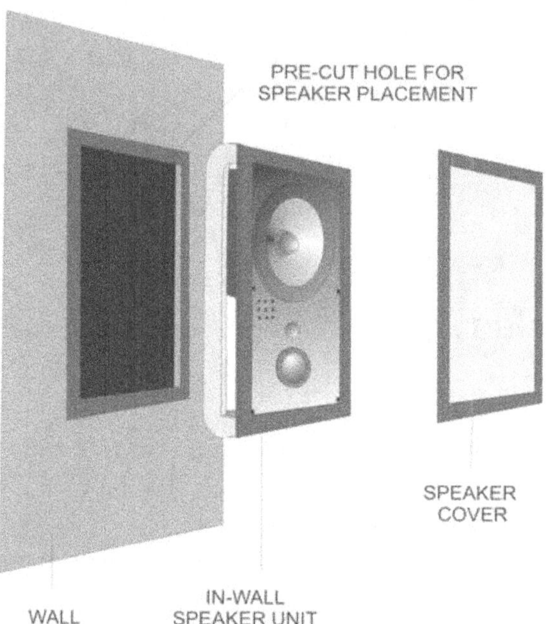

**Figure 1-17:
In-Wall Speakers**

Modern ceiling-mounted loudspeakers are capable of providing timbre-matched overhead sound solutions for all residential environments. They are often fully-pivoting.

The traditional downward coverage of fixed ceiling loudspeakers makes their placement an important consideration during the design stage of residential A/V installation.

Recent developments in home theater speaker technology include speaker assemblies for woofers and midrange units that can pivot up to 20 degrees. Rather than the typical downward coverage of traditional ceiling loudspeakers, this animated design permits sound to be directed towards the intended listening area.

Although in-wall speakers offer several advantages they must comply with certain building code requirements. Due to building safety standards, special UL/CL3-certified speaker wire must be used for in-wall installations. More information on installing and selecting the proper speaker cable is given later.

In-wall speakers offer good quality sound, and with proper installation they blend well into the room's decor. In-wall speakers conserve floor space and provide the flexibility of having the sound system in any room you prefer.

Bookshelf Speakers

Bookshelf speakers, as shown in Figure 1-18, are a good choice where space is limited. They offer excellent performance in small enclosures. They typically come in two-way systems with a crossover network. They produce good sound at the medium- and high-frequency ranges but due to their limited size are not able to produce the lower bass frequencies as effectively as floor-standing models. A powered subwoofer can extend performance in the bass frequency range.

Figure 1-18: Bookshelf Speakers

Floor-Standing Speakers

Floor-standing loudspeakers, as shown in Figure 1-19, have been around for many years. They are considered by many audio engineers to be the standard against which all other speaker types are rated. They are very efficient, and can be used in high-power applications.

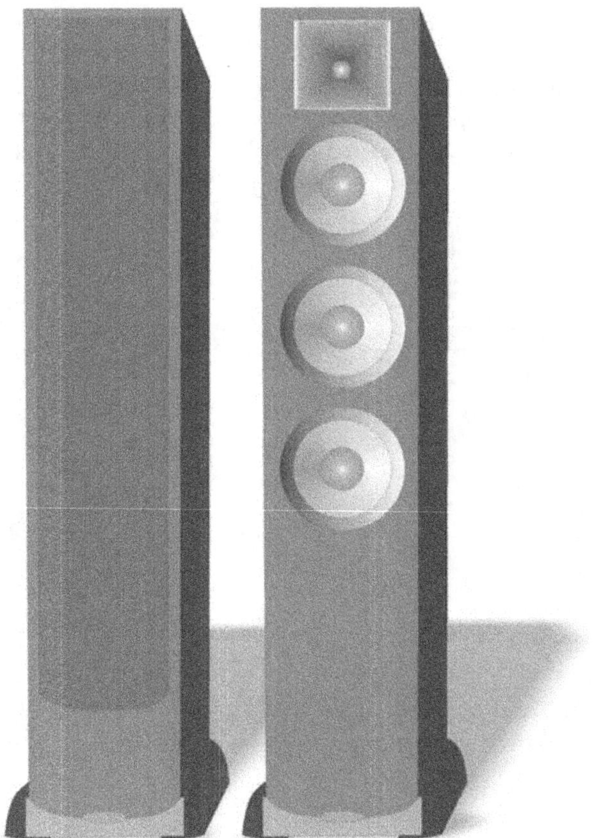

Figure 1-19: Floor-Standing Speakers

Speaker Enclosure Types

Cone speakers are mounted in various types of enclosures called **baffles** to improve the quality of the sound. A wall can often serve as baffle. Baffle types are open and closed, bass-reflex, which uses an open port, and a dual-cone baffle enclosure.

The driver of any speaker is a precision instrument. Constantly changing signals and parts that move thousands of times per second—we need to make sure the driver is protected from anything that may interfere with its operation.

The other concern with drivers is the way in which air is moved. We already know that when a driver pushes out, air is pushed out as well. The same is true when the driver is pulled back in - air gets moved in the opposite direction. The air that moves back behind the driver has to go somewhere and often times will interfere with the air being pushed forward, especially if it is reflected against a wall.

To counter the issues of protection and interference, a **cabinet** (or **enclosure**) is used. Not only does this structure provide a secured mounting surface for the driver, it also keeps the air from the front of the driver from interfering with the air in the back.

There are two main types of cabinets:

- Sealed enclosures

- Ported (Base Reflex) enclosures

The **sealed cabinet**, like the one depicted in Figure 1-20, is kept airtight to keep the driver from moving when it is not supposed to. Constant air pressure in the cabinet keeps the driver at rest. While this is usually results in a smaller cabinet size and more accurate playback, the air pressure in the cabinet often requires a higher power level be used.

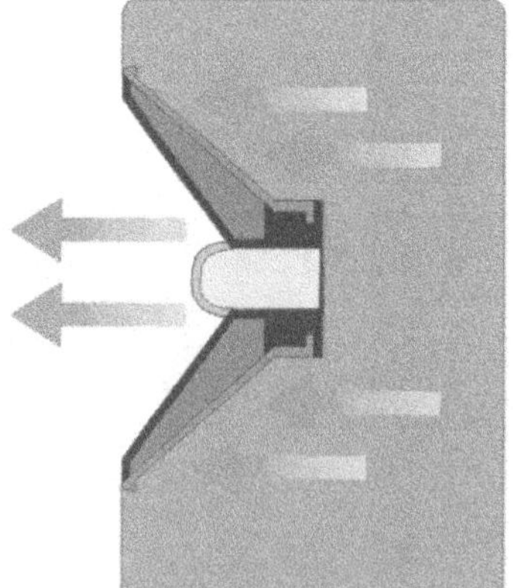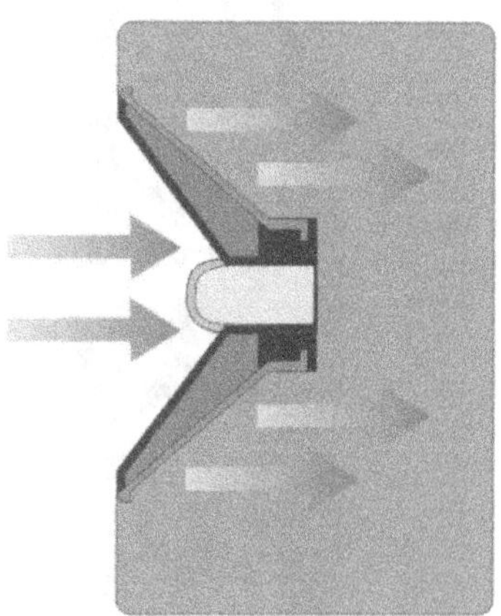

Figure 1-20:
Sealed Enclosure

Also known as a **ported design**, this term covers many design variations. They all have some kind of opening to allow increased air movement which results in more efficient performance.

Ported enclosures, as illustrated in Figure 1-21, are more complicated to design and construct, and are not as uniform in how sounds are played back. However, their increased efficiency results in a higher output level, and they can be built to **target** a certain frequency range that needs to be emphasized.

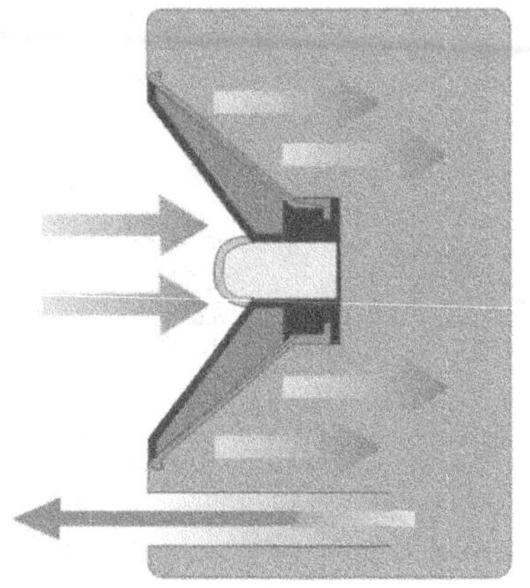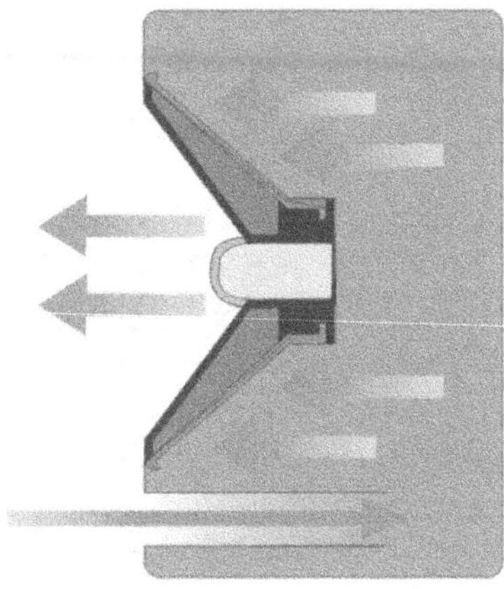

Figure 1-21: A Ported (Base Reflex) Enclosure

Speaker Crossover Networks

Modern speaker enclosures usually contain more than one speaker. Multiple speakers produce quality sound over the full range of the audio spectrum using a crossover network.

The crossover network circuit inside the speaker enclosure divides the audio signal into various frequency bands before sending them to the different speaker drivers.

Crossover networks can often be used in three-way systems for the high-frequency, mid-range, and low-frequency speaker systems. Two-way systems may employ a high-range and a low-range system.

The basic components of crossover networks are **inductors** and **capacitors**. Inductors become more reactive (increasing AC resistance) as the frequency increases. Inductors therefore block most of the high frequencies while allowing the lower frequencies to reach the low-frequency speaker referred to as a **woofer**. This portion of the crossover circuit is called a "low-pass" filter, which indicates that the lower frequencies are *passed* to the woofer.

Capacitors work just the opposite from reactors. They have lower reactance (AC resistance) as the frequency increases; therefore they pass the high frequencies to the high-frequency speaker called a tweeter while blocking the lower frequencies.

This portion of the crossover circuit is called a "high-pass" filter, which means the high frequencies are passed to the tweeter. The point where the two frequency response curves cross is appropriately called the **crossover point**. A simple two-way speaker system with a crossover point of 1000 hertz is shown in Figure 1-22. The curves for the low-pass and high-pass filters show the response on each side of the 1000-hertz crossover point.

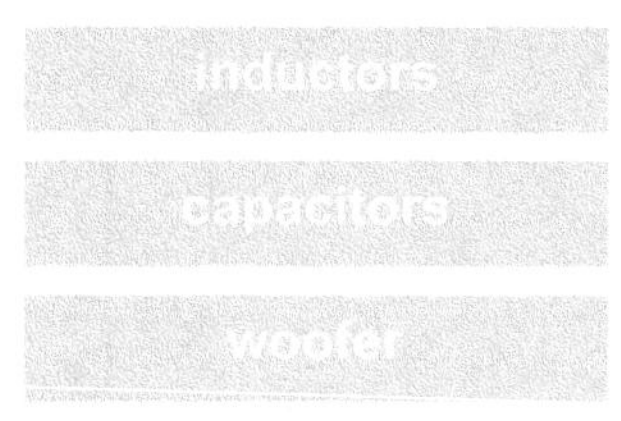

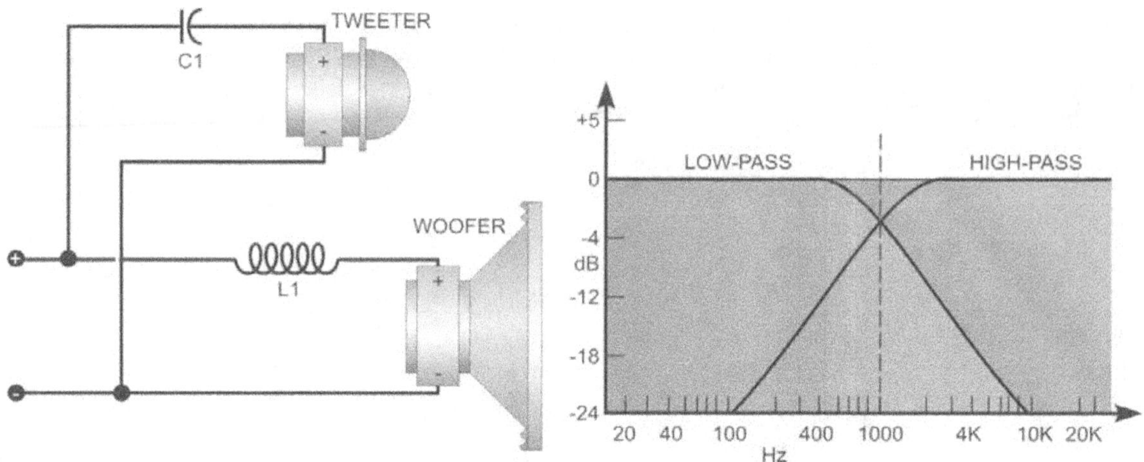

Figure 1-22: Two-Way Crossover Network and Speaker System

A **bandpass filter** can be used to select a midrange frequency band by combining a low-pass and a high-pass filter with the appropriate inductor and capacitor values. The values of the low-pass, midrange, and high-pass filters can be adjusted for crossover points to match the type of speaker used.

Figure 1-23 shows a three-way crossover network and speaker system with crossover frequencies of 500 and 5000 hertz. This type of crossover network is the basic design used with most three-way speaker systems.

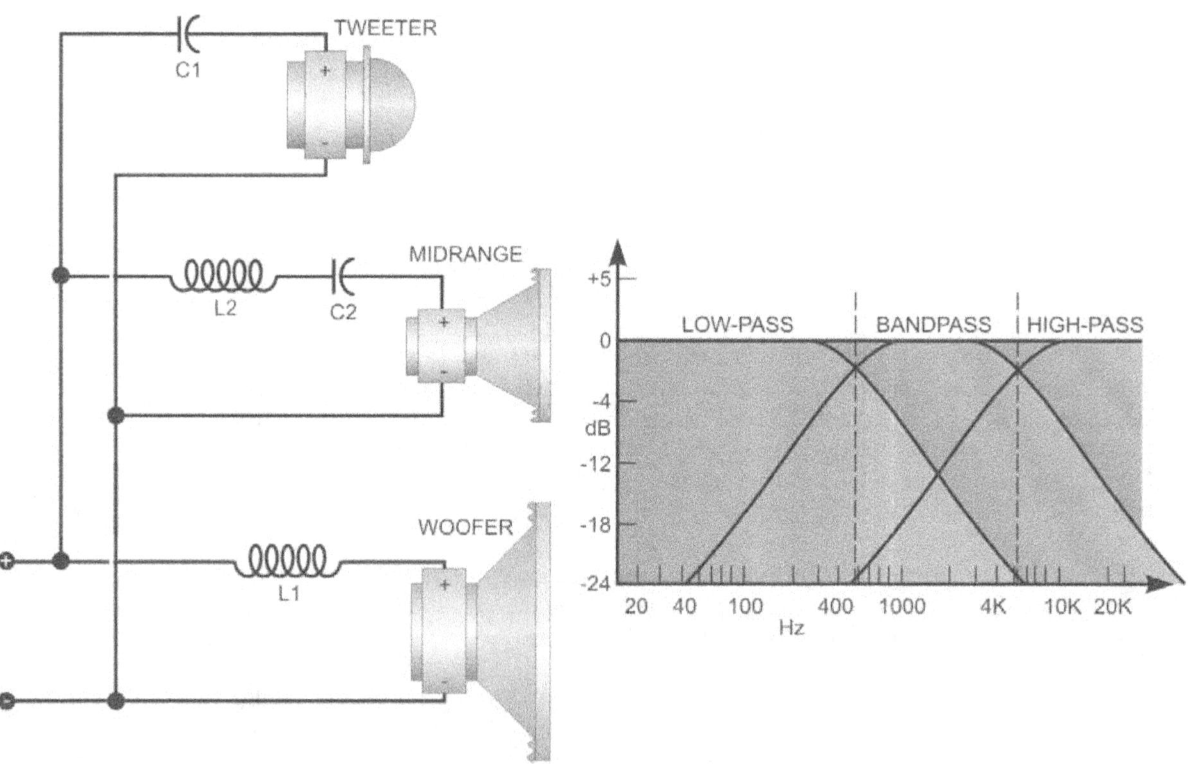

Figure 1-23: Three-Way Crossover Network and Speaker System

Speaker Power Rating

The speaker's **power rating** is a specification that indicates how much power in watts the speakers can handle without damage.

Speakers are more likely to be damaged by an underpowered amplifier than a high-power amplifier due to "clipping" with an overdriven low-power amplifier. This causes distortion of the sound and may cause the amplifier or speaker to fail. When an amplifier is too small for the speaker system, the amplifier may automatically shut down at high volume levels due to clipping of the output audio signal. This problem can be cured by using a higher-powered amplifier consistent with the power rating of the speakers.

Speaker Impedance

Like all electrical devices, a speaker creates resistance based on the work it does. Consider a vehicle towing a trailer - it needs to provide more power as the trailer gets heavier. If more power is not provided, the extra weight will slow down the vehicle.

In the audio world, **impedance** is the term used to describe the resistance a source device such as an amplifier will encounter when trying to drive a speaker or speaker system. Speaker impedance affects amplifier output. As the impedance of the load increases, the wattage output from the amplifier decreases. Conversely, when the impedance drops, wattage output from the amplifier increases. Impedance that is too low or too high can lead to distortion that can damage an amplifier.

To achieve the maximum power transfer to the speaker(s) you must match the impedance of the load (speakers) being driven with the output of the driving device. This is referred to as **impedance matching** and is the rule for connecting speakers in A/V systems.

Speaker impedance is measured in **ohms**. Home theater speakers are usually rated at 8-ohms, but are also found in four and six ohm varieties.

When multiple speakers are connected to a single output in different configurations, the resulting combinations will have different ohm ratings. You may wire speakers in parallel with each other, or in series with each other, or in a series-parallel combination to cover a large area. Again, the impedance of the speaker network must equal the output impedance of the driving output. A safer, easier method of connecting multiple speakers to an output is to place an **impedance matching volume control** between the output and the speaker network. The volume control device prevents damage from occurring to the driving device.

Amplification and Speaker Impedance

Ohm's law is used to show the relationship between voltage, current and resistance in an electrical circuit:

Ohm's Law: $V = I \times R$ where V = voltage, I = current, R = resistance.

However, a loudspeaker isn't a simple resistance because it is an electro acoustical-mechanical device, which is usually governed by a complex passive crossover network comprised of inductors, capacitors and resistors. Thus the speaker system presents complex impedance, which varies with frequency and power level

With reference to impedance, Ohm's Law can be restated as follows:

In an electrical circuit, current flow is directly proportional to voltage and inversely proportional to impedance. Mathematically, this becomes—*Current* (in amperes) *equals Voltage* (in volts) *divided by Impedance* (in ohms).

$$\textbf{Current} = \frac{\textbf{Voltage}}{\textbf{Impedance}}$$

As an example, if an amplifier is producing 10 volts AC to an 8 ohm speaker, the current in the speaker will be 10 volts / 8 ohms or 1.25 amperes. If the amplifier output is increased to 20 volts to that 8 ohm speaker, the current becomes 20 Volts / 8 ohms or 2.5 amperes. So increasing the voltage increased the current. If the voltage decreases back to 10 volts, the current will decrease back to 1.25 amperes.

Now, if our amplifier with 10 volts output is connected to a 4 ohm speaker, the lower impedance will allow more current to flow. The amount will be found by 10 volts / 4 ohms = 2.5 amperes. If we use a 2 ohm speaker, even more current flows: 10V/2 ohms = 5 amperes

Finally, if we can measure or in some other way determine the amount of current being drawn from the amplifier, we can calculate the value of the load impedance using Ohm's Law. The formula for this is: *Impedance* (in ohms) *equals Voltage* (in volts) *divided by Current* (in amperes).

$$\textbf{Impedance} = \frac{\textbf{Voltage}}{\textbf{Current}}$$

Power is measured in watts and is equal to:

$$P = E \times I$$

where P is power, E is voltage, and I is current. The equation can be used to solve for E with the formula $E = P/I$ and $I = P/E$.

Watts vs. the Decibel (dB)

A **watt** is used to describe the energy output of a receiver or amplifier used to power a loudspeaker. The relationship between power output and speaker loudness or volume is not linear or straight (+10 watts does not equal +10 dB).

Decibels are expressed as a ratio of two powers. As an example, dBs = 10log (P2 P1) where P1 and P2 are the two powers being compared and where the log is to the base 10. The **Decibel** was invented to represent sound levels but has found many other uses wherever a ratio or comparison of widely differing values is required. Because the Decibel is logarithmic small changes in value represent large differences. It must be remembered that the decibel does not have a unit (being a simple ratio). Therefore, when it is used, frequently a unit is associated with it. It can then be read as X Decibels relative to the unit that is used.

The amount of loss in sound pressure level (SPL) as you double the distance away from a sound source is approximately 6 dB.

The most frequent unit used in comparisons is the unit of power which is measured in Watts. Because powers vary so much in normal day to day electronics another unit relative to 1 milliwatt or 1×10^{-3} Watts is also frequently used, this is written as dBm. A signal with a power of 0 dBW is exactly the same as a signal with 1 Watt. Table 1-1 provides a good way to get an idea of why Decibels are so useful.

POWER IN WATTS	POWER IN dBm	POWER IN dBW
1	30	0
100	50	20
1,000	60	30
1,000,000	90	60
0.01	10	-20
0.001	0	-30
0.000001	-30	-60
0.000000001	-60	-90
0.000000000001	-90	-120

Table 1-1: Comparison of Power Ratios Using Decibels

An increase of 3 dB represents a doubling of power. The reverse is also true. If a signal decreases by 3 dB, half the power is lost. When a 1,000-watt signal decreases by 3 dB, its value stands at 500 watts, while a 1,000-watt signal that increases by 3 dB will assume a value of 2,000 watts.

Keep in mind that the term *local amplification* refers to a distribution format featuring multiple speaker systems that receive their power from lower power localized power amplifiers. Alternately, *centralized amplification* refers to a distribution format featuring a single power amplifier capable of driving multiple speaker systems from one location.

Audio Wire

Audio wire is used to connect sound equipment in home theater systems. Three types of cable are required for audio connectivity. Coaxial shielded cable and optoelectric (fiber-optic patch cord) are used for audio component connectivity, and speaker wire is used to connect amplifier output terminals to speaker systems. Coaxial shielded cable and optoelectric (fiber-optic patch cord) are used for audio component connectivity, "at line levels and below." Speaker wire is used to connect "speaker level" amplifier output terminals to speaker systems.

Typical types of audio cable used for low-voltage structured cable connectivity are described as follows:

- *Shielded twisted pair* – This audio cable is commonly used to carry balanced microphone signals and balanced line level signals. Balanced connections and good STP cables are needed to transfer audio signals over relatively long distances.

- *Shielded single conductor* – This cable is used to carry unbalanced audio signals. It has one central coaxial signal conductor and a ground shield. This cable type is sometimes called *high-impedance cable* because it typically is used with high-output impedance equipments such as tape and CD players and musical instruments (such as an electric guitar).

- *Unshielded twisted pair cable* – This cable, usually Category 5e, is used in telephone (voice) and home network wiring (structured cabling systems). It is not very suitable for professional audio use. With suitable adapters, Category 5e cable can carry digital and analog audio signals, but with limited performance.

- *75-ohm coaxial cable* – This cable is used to carry digital audio signals between audio equipment that requires a 75-ohm unbalanced connection.

- *Toslink cable* – This is a type of fiber-optic cable that transfers audio information in a digital format rather than an analog format. Toslink cable can be used with all audio components equipped with a fiber-optic connector.

Standard Speaker Wire

TEST TIP

Be familiar with the special requirements of speaker cable for outdoor use.

The most common type of audio connection is the basic **speaker wire**. Each speaker requires two lines, or conductors. Since a speaker is an electrical device, it requires a positive and a negative line for a current to make a complete circuit. These conductors can either be made up of many smaller wires stranded together, or a single solid conductor. While there are debates over the benefits of *stranded* versus *solid*, they both have the same function of taking a current from an amplifier, and allowing it to pass to the speaker.

Wires are wrapped in an insulating layer called the **sheath** or **jacket** which separates and protects the conductors. Typically, a 2-conductor wire will handle a single speaker, 4-conductor will handle 2 speakers, and so on.

The thickness of the conductors also matters—the longer the distance, the thicker the wire needs to be. The wire also needs to be thicker as the amount of power being used increases.

The thickness of the wire is determined using the **American Wire Gauge (AWG)**. This specification refers to how many strands were originally used to form a certain size wire—a larger wire would require less stranding, which is why the number goes down as the wire diameter increases.

In general:

- Standard speaker wire is available in the range of 12 to 16 gauge.

- Thicker wire is able to pass the audio signal from the amplifier output to the speaker with less resistance.

- The general rule for installing speakers is to limit the distance between the amplifier and speaker to a maximum of 50 feet.

- A minimum of 14 gauge wire should be used for runs of 100 feet.

- The minimum gauge speaker wire recommended for any location is 16 gauge.

VIDEO DISPLAY DESIGN CONCEPTS

Like receivers, picture devices are made up of a few common components:

- **Display** — It may seem obvious that a picture device has a component to display an image, but there are many different types and technologies. While we will explore the different technologies further, the basic types are as follows:

    - **Reflective** — This type of display uses a mirror or other mechanical advantage to increase the image size without building a full size display panel, as illustrated in Figure 1-24. DLP and other projection devices fall into this group.

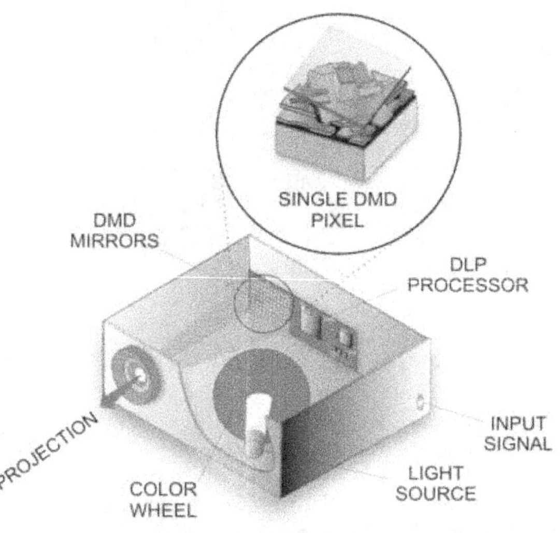

DLP PROCESSING

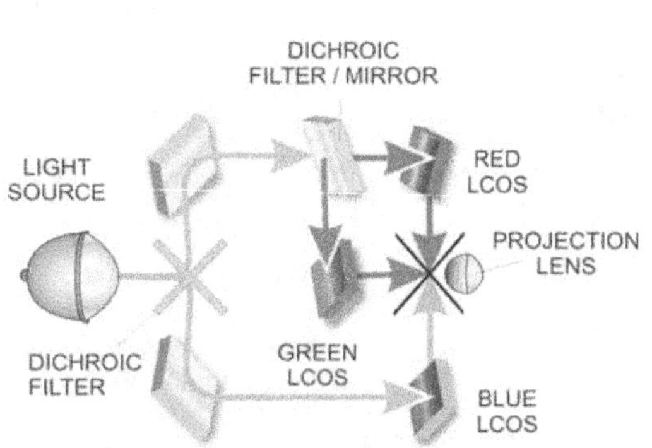

LCOS OPERATIONS

**Figure 1-24:
Reflective Displays**

- **Emissive** − This is any display device that emits the image directly to the viewer, as depicted in Figure 1-25. It does not pass through any materials; the viewer sees the image as it is presented. This includes plasma and Cathode Ray Tube (CRT) displays.

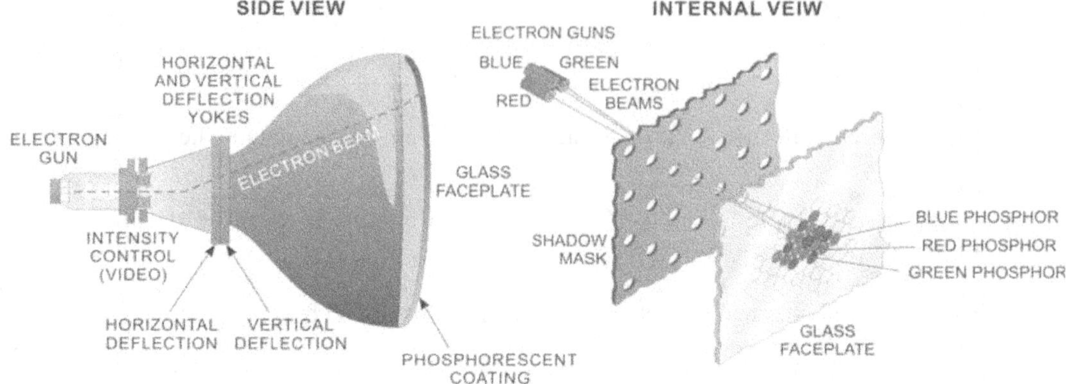

**Figure 1-25:
Emissive Displays**

- **Transmissive** − This technology has the image (or it's light source) pass through another material. The most common example would be LCD based technology that uses a backlight, as shown in Figure 1-26.

**Figure 1-26:
Transmissive
Displays**

- **Tuner** − We covered the different signal types in AV Theory, but there are two main differences between a television and a monitor. A television has the ability to decode channels from an external cable or antenna feed, whereas a monitor can only display signals that have been decoded by an external source.

- **Speakers** − Another difference between a television and a monitor is that televisions usually have built in speakers. TV speakers use small drivers, with limited power and poor sound compared to separate sound systems. In many cases, they must be disabled when sound is being sent from the display to an external audio device.

Cathode Ray Tube Displays

Cathode Ray Tubes (CRTs) have been used as the main display component of television receivers and desktop computer monitors for several years. Although popular and relatively inexpensive, they are limited by economical and technology constraints to display areas of approximately 36 inches. CRT monitors and TV displays are specified in size by the diagonal measurement of the tube's front area.

CRTs are classified according to the number of **pixels** (picture elements) that they can display. On a color display, each pixel is composed of a red, a blue, and a green dot.

The number of pixels that can be displayed on a CRT display is called the **display resolution**. As an example, a VGA computer monitor has a resolution of 640 x 480 pixels and a SVGA monitor has a resolution of 800 x 600 pixels.

The design features of a CRT are illustrated in Figure 1-27.

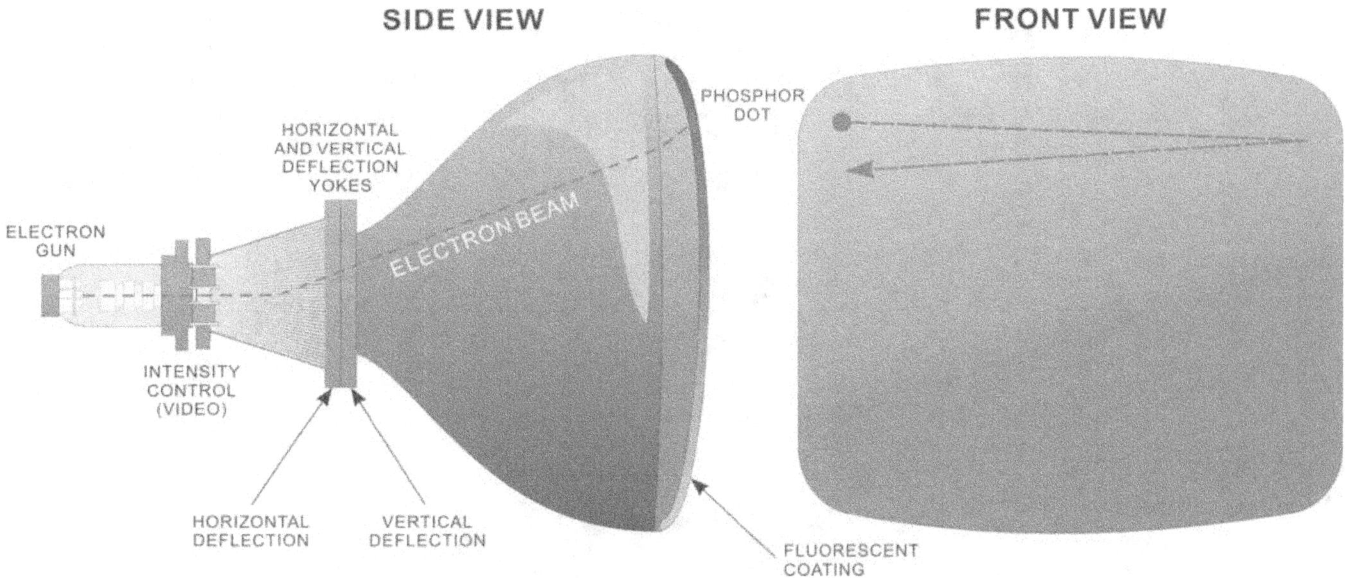

Figure 1-27: Cathode Ray Tube Display

Flat-Panel Display Technologies

Flat-panel displays get their name from the physical shape and appearance that distinguishes them from CRT displays. They are suitable for lightweight portable laptop computers and new large-screen high-definition TV (HDTV) receivers. The leading technologies for flat-panel displays are liquid crystal and plasma.

Remember that a CRT is not a flat-panel display. As mentioned earlier, some CRT monitors and TVs using a flat-faced CRT design are often listed in vendor specifications as flat-screen monitors and TVs.

Liquid Crystal Displays

Liquid crystal displays (LCDs)

Liquid crystal displays (LCDs) are the most common type of flat-panel display. They are used for a wide range of small portable electronic and computing devices including laptop computers, cell phones, games, and personal digital assistants (PDAs). LCD displays utilize two sheets of polarizing material with a liquid crystal solution between them, as shown in Figure 1-28. An electric current passed through the liquid causes the crystals to align so that light cannot pass through them. Each crystal, therefore, is like a shutter, either allowing light to pass through or blocking the light, as shown in Figure 1-29.

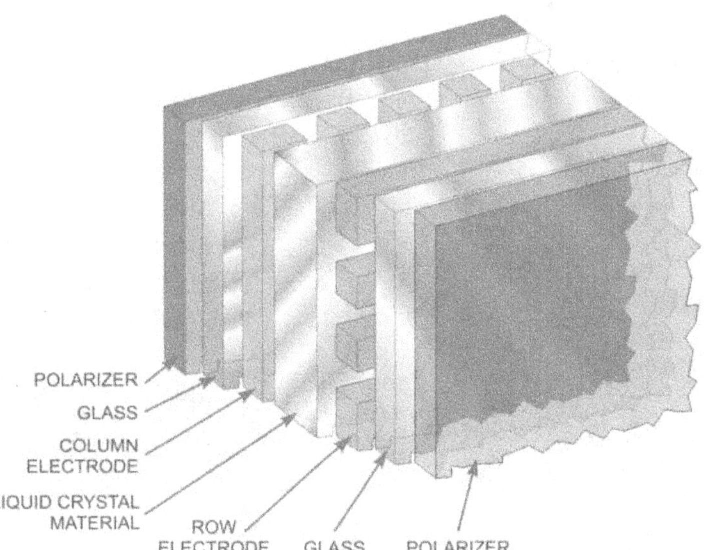

**Figure 1-28:
LCD Construction**

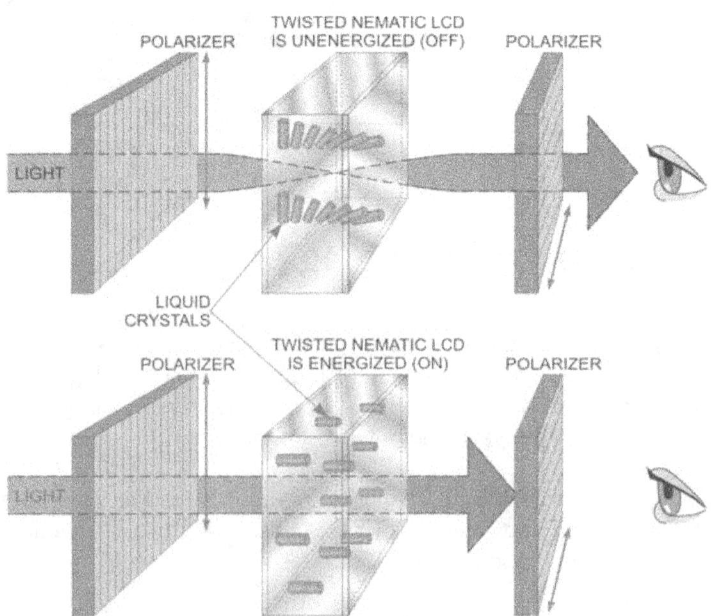

**Figure 1-29:
LCD Operation**

Color LCD displays are created by adding a three-color filter to the panel. Each pixel in the display corresponds to a red, blue, or green dot on the filter. Activating a pixel behind a blue dot on the filter will produce a blue dot onscreen. Like color CRT displays, the color on the screen of the color LCD panel is established by controlling the relative intensities of a three-dot (RGB) pixel cluster.

A display screen made with TFT (thin-film transistor) technology is an advanced liquid crystal display design, common in top-of-the-line laptop computers, that has a transistor for each pixel. TFT is also known as "active matrix" display technology and contrasts with "passive matrix" LCD technology, which does not have a transistor at each pixel. An active matrix display can be switched on and off very rapidly, presenting a fast, high-resolution appearance. Active matrix is the superior technology for LCD displays.

Plasma Display Panels

Plasma display panels are a further improvement in flat-panel display technology. A plasma display is a type of advanced large-screen TV display.

Plasma display panels contain hundreds of thousands of tiny cells (pixels) containing minute amounts of an inert gas sandwiched between two sheets of glass, as shown in Figure 1-30. Electrodes are placed in pairs on the inner side of the front plate. When electrically charged, they produce an ultraviolet beam, which activates the phosphorous coating of the cell and transmits light through the glass surface.

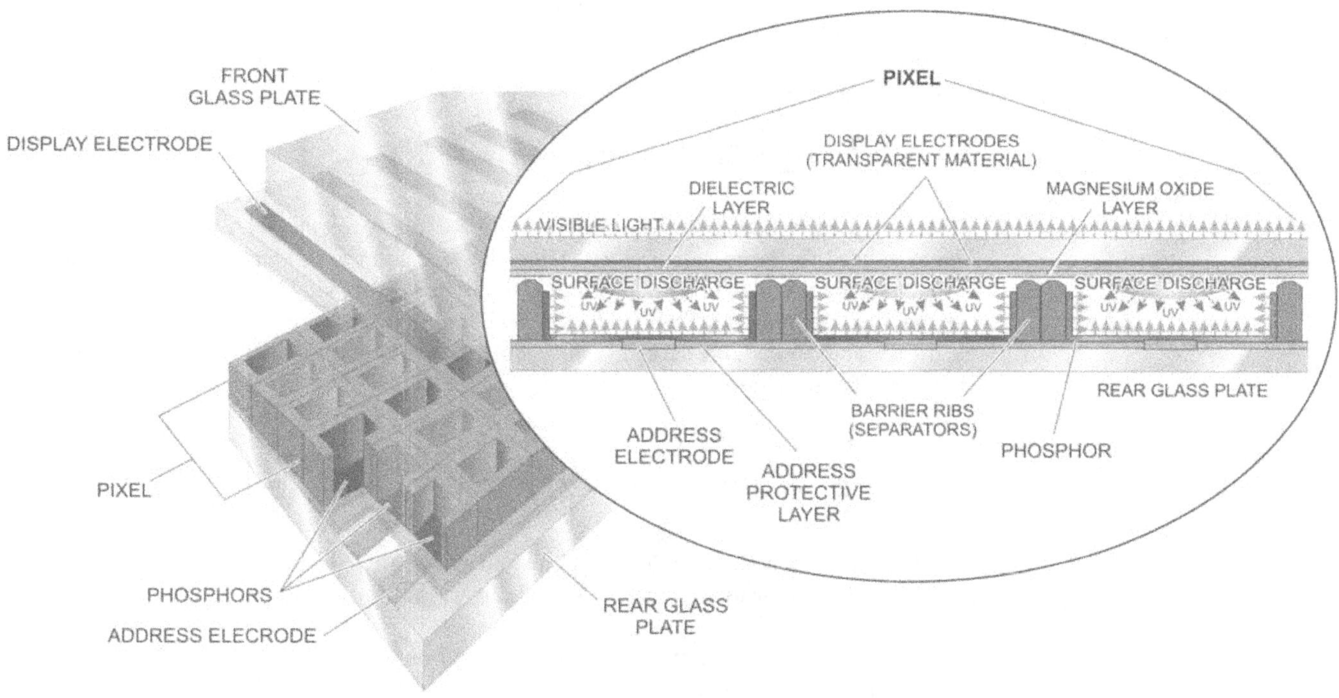

Figure 1-30: Plasma Display Components

Plasma displays are available for large-screen TV displays offering high resolution and a 16:9 aspect ratio compatible with the HDTV format. The flat-panel design allows them to be wall mounted or in component cabinets.

Plasma displays are expensive and range in screen size from 42 to 50 inches with a thickness of only 4 inches.

Burn-in and Plasma Displays

Burn-in, sometimes also referred as permanent image retention, is an effect of an after-image appearing on plasma or other phosphor-based screen after a still picture is displayed for an extended period of time. The most common type of burn-in is the result of watching 4:3 video with black bars on both sides on a 16:9 screen. Plasma TVs are simply not designed to display fixed images for extended periods of time. Burn-in occurs because of uneven wear of phosphors on the display.

Video Projection Systems

Video projection uses a high-intensity light source to project video images on a screen. They are used for large-screen TV displays as well as multimedia meeting presentations. Unlike plasma, LCD, and CRT displays, which are "direct view" technologies, projection systems are "indirect view" video displays where the image size is enhanced by projecting the video image on a large screen. Video projection is accomplished using either front projection or rear projection.

Front Projection Video

Front projection utilizes a video projector and a separate pull-down screen to show the projected image, similar to a movie screen, as shown in Figure 1-31.

The video projectors are often ceiling-mounted for home theater installations. High-end front projection systems can achieve an image size of 3 to 25 feet. Front projection systems are used for computer-based presentations as well as projection TV systems for large rooms or auditoriums.

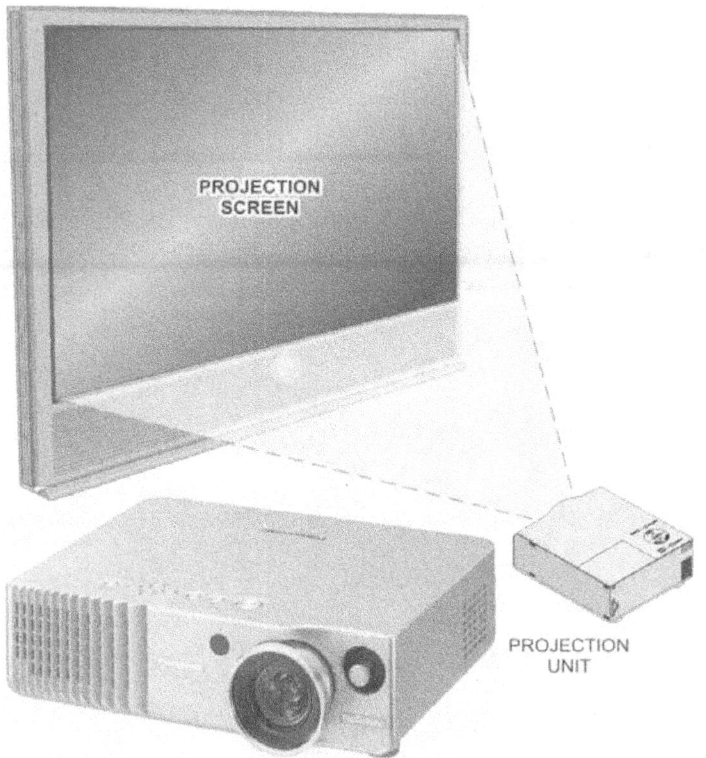

Figure 1-31: Front Projection TV System

Rear Projection Video

Rear video projection combines the projector and the viewing screen into one television unit. A typical **rear projection** unit is shown in Figure 1-32.

A rear projection TV display contains an integrated three-gun (red, green, and blue) projection system that is aimed at a mirror. A colored image is formed by the three picture tubes and is projected onto a mirror. The image is then reflected onto the rear of a display screen. The mirror enables the image to travel a sufficient distance between the projector and the screen to provide an image of between 40 to 80 inches.

Rear projection systems are a popular choice for home theater systems where room size prohibits the use of a front projection system.

TEST TIP

Know the advantages offered by rear projection TV display systems.

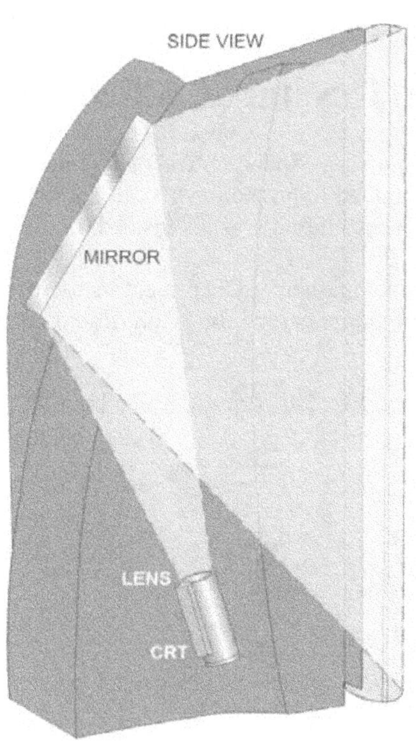

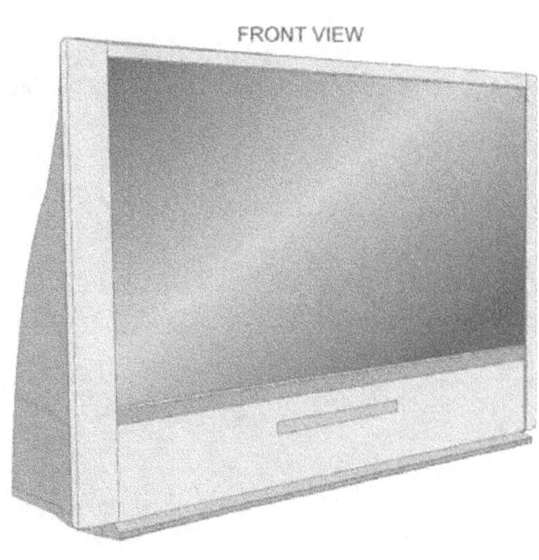

Figure 1-32: Rear Projection TV

Digital Light Processing

Digital Light Processing (**DLP**) is a new display technology. The DLP trademark is owned by Texas Instruments, representing a technology used in projectors and video projectors. DLP processing is shown in Figure 1-33.

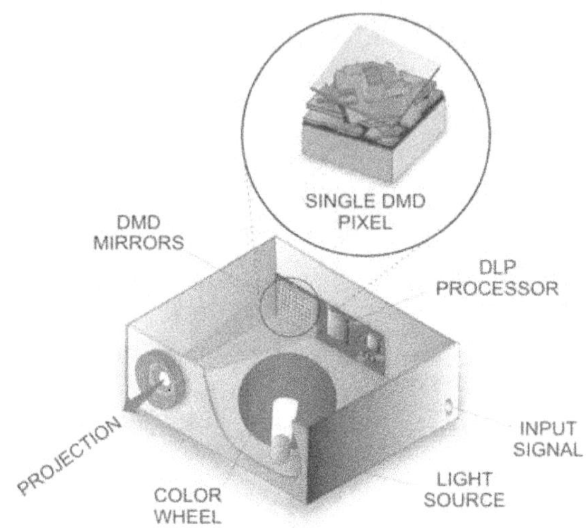

**Figure 1-33:
DLP Processing**

One application is DLP front projectors (small standalone projection units). DLP, along with LCD and **Liquid Crystal on Silicon** (**LCoS**) are the current display technologies behind rear-projection television, having for the most part, supplanted CRT projectors. These rear-projection technologies compete against LCD and Plasma flat panel displays in the **High Definition Television** (**HDTV**) market.

Liquid Crystal on Silicon (LCOS or LCoS)

Liquid Crystal on Silicon (LCoS) is a "micro-projection" or "micro-display" technology typically applied in projection TV. It is a reflective technology similar to DLP projectors; however, it uses liquid crystals instead of individual mirrors.

By way of comparison, LCD projectors use transmissive LCD chips, allowing light to pass through the liquid crystal. In LCoS, liquid crystals are applied directly to the surface of a silicon chip coated with an aluminized layer, with some type of passivation layer, which is highly reflective. LCoS operations are shown in Figure 1-34.

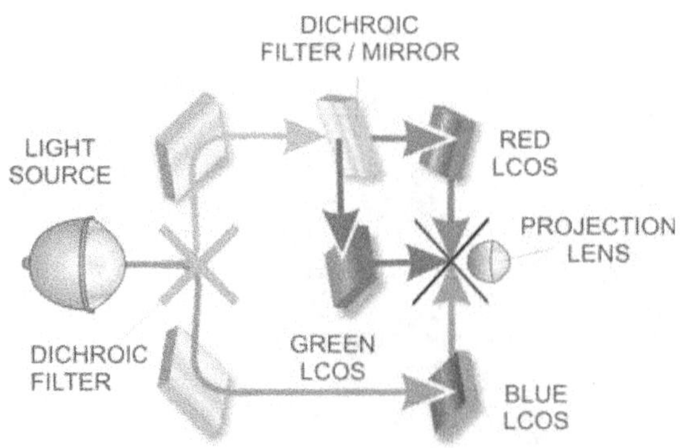

**Figure 1-34:
LCoS Operations**

ANALOG AND DIGITAL TELEVISION BROADCAST STANDARDS

Commercial television broadcasting is in a transition between the gradual phase-out of 50-year-old analog broadcast standards and new digital television transmission. The **Federal Communications Commission (FCC)** has mandated that all stations begin broadcasting digital **High-Definition TV (HDTV)** on February 17, 2009. This section will prepare you for understanding both existing analog and future digital television broadcast standards that will have a major influence on the home entertainment market as well as future opportunities for the home technology integrator.

Analog Television Transmission Standards

The three existing analog world television broadcast standards in use today are:

- National Television System Committee (NTSC)
- Phase Alternating Line (PAL) (Europe)
- Sequential Electronique Couleur Avec Memoire (SECAM) (France)

The National Television System Committee

The **National Television System Committee (NTSC)** developed the standard as well as the name for transmission of television signals in the United States. It prescribes a vertical resolution of 525 lines on the TV tube (the cathode ray tube). The lines are stacked from the top to the bottom.

The field rate is set at 60 fields displayed per second. A field is defined as a set of even lines or odd lines. The odd and even fields are therefore transmitted to the TV set sequentially. All odd and even sets of lines are transmitted and interlaced on the screen producing a full frame period of interlaced lines in approximately 1/30 of a second, as illustrated in Figure 1-35. This technique is known as **interlaced scanning**. The NTSC system is called a 30-frames-per-second system. NTSC is also used as the standard for recording television signals on tape such as VHS. NTSC is used in the United States and several other countries.

TEST TIP

Be aware that the analog commercial television transmission standard for television in the United States is the National Television System Committee (NTSC) standard. It is a 525-line 30-frames-per-second interlaced standard. PAL and SECAM are transmission standards used in Europe and France, respectively.

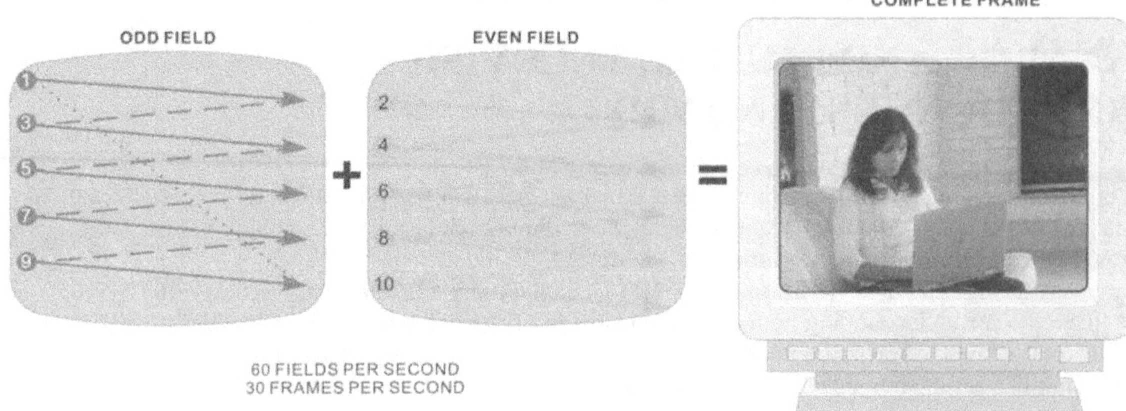

**Figure 1-35:
Interlaced
Scanning**

Keep in mind that video formats used for consumer electronic products are classified according to the number of horizontal lines of resolution contained in each frame. A higher number of lines produce a progressively higher quality picture.

Phase Alternation by Line

Phase Alternation by Line (PAL) is the television transmission standard used in several Europe and Asian countries. It has a vertical resolution of 625 lines on the TV tube and a field rate of 50 fields per second. Using a similar interlace scheme as the NTSC standard, PAL produces a frame rate of 25 frames per second.

SECAM

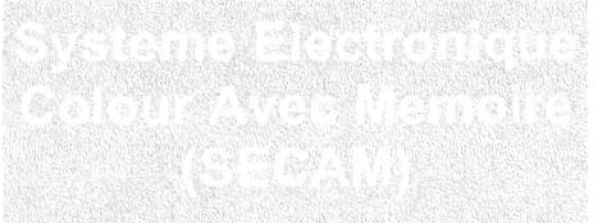

Systeme Electronique Colour Avec Memoire (SECAM) is a transmission standard used in France. It also has a 625-line vertical resolution and a 25-frames-per-second frame rate but transmits the color information differently from the PAL system.

In radio communications, a **vestigial sideband (VSB)** is one that has been partially cut off, or suppressed. Television broadcasts (in NTSC, PAL, or SECAM analog video format) use this method if the video is transmitted in AM, because of the large bandwidth required. In digital transmissions it is used in ATSC as standardized 8-VSB. In countries that use NTSC or ATSC, the video baseband signal for television requires a bandwidth of 6 MHz. The use of Single Sideband (SSB) would conserve bandwidth, however, the video signal contains significant low frequency content of average brightness, along with the rectangular synchronizing pulses.

Vestigial sideband modulation offers an acceptable compromise is. In normal vestigial sideband transmissions, the full upper sideband bandwidth is 4 MHz, while the lower transmitted sideband is only 1.25 MHz. Of course the actual transmission also includes a carrier, making the system effectively AM at lower frequencies and SSB at higher frequencies. RF and IF filters help to compensate for the absence of the lower sideband components at high frequencies.

Digital Television Broadcast Standards

Digital television broadcasting standards have been established for the United States commercial broadcast industry. The following paragraphs discuss the features and specifications collectively described as **Digital TV (DTV)**.

Advanced Television Systems Committee and DTV

Digital TV is a new "over-the-air" digital television standard that will be used by nearly 1600 local broadcast television stations in the United States. The DTV standard was incorporated into the Federal Communications Commission (FCC) rules as ATSC Document A/53 (ATSC Digital Television Standard).

The DTV standard will allow commercial broadcasters to provide higher quality services. First, DTV will permit transmission of television programming in new wide-screen, high-resolution formats known as high-definition television (HDTV). In addition, the new DTV television system allows transmissions in standard-definition television (SDTV) formats that provide picture resolution similar to existing television service. Both the HDTV and SDTV formats will have significantly better color rendition than the existing analog television system. None of the DTV transmissions will be able to be received on existing analog TV receivers.

Quadrature amplitude modulation (QAM) is the format by which digital cable channels are encoded and transmitted via cable. QAM tuners show up in some North American digital video TVs that enable direct reception of digital cable channels without the use of a set-top box, permitting analog cable subscribers to receive QAM-based HD programming of local stations, without having to pay for a digital cable box. However, the availability of QAM HD programming is not touted or publicized in the product literature of cable companies. QAM tuners are cable equivalents of ATSC tuners required to receive over-the-air (OTA) local digital channels broadcast. Newer cable-ready digital televisions support both ATSC and QAM, which carries nearly twice the data of over-the-air ATSC 8VSB.

An integrated QAM tuner is entirely appropriate for digital cable as it requires a clean signal path. Cable providers use it to provide free reception of clear, unscrambled, and usually high definition digital programming by local broadcast stations and cable radio channels. However, the majority of digital channels use scrambled signals because the providers consider them to be extra-cost options, rather than part of the basic cable package. Digital cable-ready TVs may already be equipped with the necessary CableCARD provided by the cable provider to unscramble various protected channels. Subscribers can then tune-in all authorized digital channels, again without the use of a set-top box.

The FCC does not require that QAM tuners be included with new TV sets, as it does with the ATSC standard, which calls for the use of 8VSB modulation. QAM permits cable providers to deliver twice as much data that 8VSB does, they petitioned the FCC to allow the use of QAM instead of 8VSB. ATSC and QAM are commonly included in most receivers because the same hardware is often used for both. However, because QAM is only a modulation, it does not specify a digital data format. For ATSC digital cable television, its modulation format is based on ATSC. Although QAM is also used with **Digital Video Broadcasting-Cable (DVB-C)**, this data format is incompatible with North American receivers.

In contrast to DVB-C, **Digital Video Broadcasting-Satellite (DVB-S)**, serves every continent of the world. As the original DVB forward error coding and modulation standard for satellite television since 1995, its transport stream is mandated as MPEG-2 using the 11/12 GHz frequency band. There is also **Digital Video Broadcasting – Terrestrial (DVB-T)**, the DVB European-based consortium standard for the DVB digital terrestrial television. It also uses an MPEG transport stream for compressed digital audio, video, and other data, with **Orthogonal Frequency-Division Multiplexing (OFDM)** modulation. OFDM splits the digital data stream into a large number of slower digital streams, instead of transporting data on a single radio frequency carrier. Each stream consists of a digitally modulated set of closely spaced adjacent carrier frequencies. DVB-T offers two choices for the number of carriers being used: either 2K or 8K. There are actually 1705 carriers used for the 2K carrier system, while the 8K carrier system utilizes 6817 carriers. The carrier spacing (4 kHz or 1 kHz) depends on the transmission channel (8MHz, 7MHz, or 6 MHz).

High-Definition TV Standards

All DTV formats are transmitted using MPEG-2 compression. There are a total of 16 DTV formats with different frame rates, interlaced and progressive scanning, and different aspect ratios. **High-Definition TV (HDTV)** uses six of the 18 formats adopted for the DTV standard by the ATSC. The remaining 12 standards are collectively identified as **Standard-Definition TV (SDTV)**.

Video processing involves a number of functions necessary to provide acceptable picture quality on consumer-level equipment, including scaling and conversions. For example, a **video scaler** is a device for converting video signals from one size or resolution to another, most often involving the up-scaling or up-conversion of a video signal from a low or standard resolution, to one of higher resolution such as in high definition television. These video scaler processors are often embedded within computer monitors, televisions, video editors, and other audio/video devices, including broadcasting equipment.

Video scalers can also be provided within a completely separate box, capable of offering simple video switching operations. Stand-alone video scaling units commonly appear in presentation and home theater systems. Within a home theater environment, a video scaler could be used to up-convert a standard definition DVD or video game signal, into a high-definition format suitable for display on a LCD or plasma television, providing a higher quality picture. Scalers find widespread application within environments where numerous video sources (DVD video, camera feeds, or DVI/VGA computer outputs) must be switched, while maintaining the highest possible resolution. Keep in mind that although video scalers are primarily digital devices, they can support analog inputs and outputs in combination with analog-to-digital converters (ADCs) or digital-to-analog converters (DACs).

Aspect Ratio

When early film technologies were being developed, the size and shape of the video information was based on the film stock and the spacing required to provide an audio track on either side, as shown in Figure 1-36.

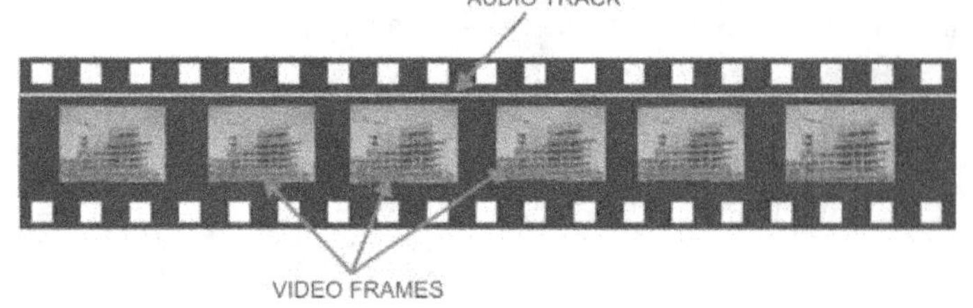

**Figure 1-36:
Film Stock**

This measurement is known as the **aspect ratio**—expressed as the ratio of width by height. The earliest aspect ratio worked out to be 4 units wide by 3 units tall—A screen that is 40 feet wide will be 30 feet tall. With the advent of television broadcasts, the same aspect ratio was used in television construction. When we examine a common computer video resolution (1024x768), we see that same ratio:

$$1024/768=1.3:1 \text{ which is also expressed as } 4x3$$

The 16:9 aspect ratio was present in the consumer electronics scene in a very limited fashion until the mid 1990's. Until that point, widescreen presentations were usually reserved for movie theaters. At this time, digital broadcasts and DVDs began to enter the mainstream consumer electronics market and the 16:9 format became the standard for HD video broadcast. Figure 1-37 shows common aspect ratios used in different types of video displays.

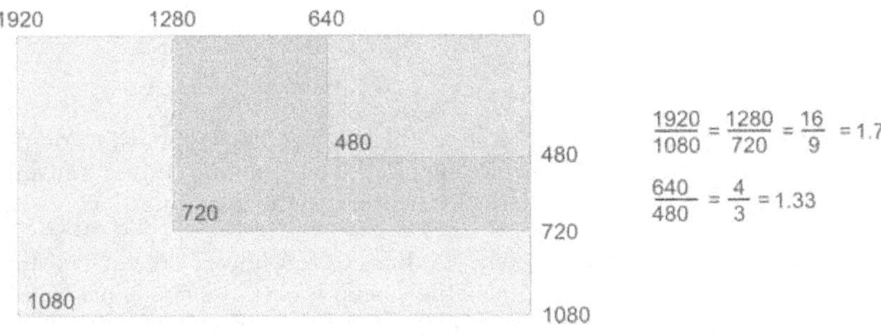

Figure 1-37: Common Aspect Ratios

The six HDTV standards with the corresponding screen resolution, aspect ratio, frames per second, and scan mode are shown in Table 1-2. The DTV standard is a transmission standard, not a display standard. Broadcasters are not forced to use an interlaced display just because it exists in the standard.

HORIZONTAL LINES	VERTICAL LINES	ASPECT RATIO	FRAMES/SEC	SCAN MODE
1920	1080	16:9	30	Interlaced
1920	1080	16:9	30	Progressive
1920	1080	16:9	24	Progressive
1280	720	16:9	60	Progressive
1280	720	16:9	30	Progressive
1280	720	16:9	24	Progressive

Table 1-2: HDTV Standards

Most broadcasters, however, have elected to use the 1080i interlaced format. It is the only one out of the six standards that uses interlaced scanning. A few broadcasters are planning tests in the future with the 720p (progressive scan) format. In a **progressive scan** format, the entire field is scanned in one continuous operation, as shown in Figure 1-38. These are the only two HDTV formats adopted at this time by commercial broadcasters.

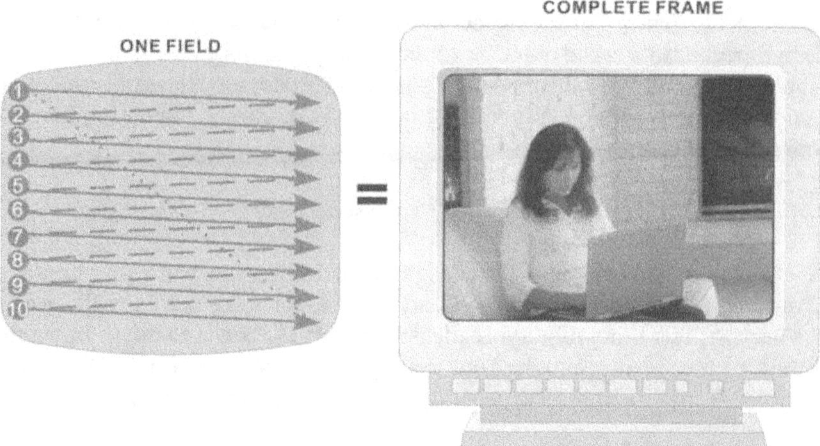

**Figure 1-38:
Progressive Scanning**

The two formats are described as follows:

- The 1080i interlaced transmission standard provides 1,080 lines of resolution (more than double the current 525 lines) using interlaced scanning. It first transmits all the odd lines on the TV screen, and then transmits all the even lines.

- The 720p progressive format offers 720 lines of resolution. This offers slightly less resolution but constructs or "paints" the screen image in a single pass. Progressive scan transmission eliminates some irregularities inherent to the interlaced display that engineers refer to as "artifacts." Due to the higher horizontal scan rate, 720p requires more transmission bandwidth than the 1080i standard and is not compatible with most existing TV broadcast transmission equipment.

> **TEST TIP**
>
> It is important to remember that the FCC policy requires commercial TV stations to broadcast all its transmissions in the HDTV format by the year 2006. Know that the HDTV transmission mode includes six of the standard transmission rates, which are a subset of the 18 DTV standards adopted by the Advanced Television Systems Committee. You need to keep in mind also that the majority of broadcasters have adopted the 1080i interlaced DTV standard for HDTV transmissions.

> **TEST TIP**
>
> Know which HDTV standard has been adopted by the majority of broadcasters.

Unlike the current NTSC TV standard, the HDTV screen has a 16:9 aspect ratio, which means that the screen is not square, but is similar to the letterbox mode occasionally used for movie telecasting.

The FCC has mandated that all stations be capable of broadcasting HDTV by 2009.

Regardless of the available screen sizes, aspect ratios, and features, clients are mainly interested in the quality of the image. A subjective analysis of image quality is still dependent on the use of industry-accepted video-quality evaluation tools, objective testing criteria, and trained experts. Among the video calibrations that receive utmost attention are color balance, and brightness (luminance), and contrast.

Color balance

Color balance is the characteristic of a multi-color image in which the relative density (brightness) of all colors is equal, with no one color dominating. In a monitor, color balance is a setting that adjusts the relative intensities of the red, green, and blue electron guns in a CRT, or the red, green, and blue pixels in a flat panel. The automatic adjustment usually makes them equal, while a manual adjustment provides a warmer display (more yellow) or cooler (more blue).

In a camera, color balance is a setting that compensates for the differences in color temperature (chromaticity) of the surrounding light. To represent all colors in the scene faithfully, the white balance must be adjusted in both film cameras and digital cameras. Preset adjustments such as "tungsten" or "fluorescent" are usually available, or automatic adjustments can be used by aiming the lens at a totally white surface, and selecting a "lock white balance" command. Alternatively, a gray card with 18% gray can be used.

This means that color balance in video production boils down to the concept of color temperature. Color temperature is a measure of the hue of a display at given brightness levels, and is expressed in degrees Kelvin. A light source with a higher color temperature (a higher Kelvin value) provides more blue light than a light source with a lower color temperature (a lower Kelvin value). Weirdly enough, a cooler light has a higher color temperature because the Kelvin scale runs opposite to the concept of measuring air temperature on a standard thermometer.

Brightness is a photometric measure of the luminous intensity per unit area of light traveling in a given direction. Describing the amount of light passing through or being emitted from a particular area, luminance characterizes light emissions or reflections from flat, diffuse surfaces that fall within a given solid angle. It indicates how much luminous power the eye will perceive, when looking at the surface from a particular angle of view. In the video industry, luminance characterizes the brightness of a video display. The video industry calls one candela per square meter a "nit", with a typical computer display emitting between 50 and 300 nits. The term is believed to originate from the Latin "nitere," meaning "to shine." Modern flat-panel (LCD and plasma) displays often exceed 300 nits.

Another video calibration related to visual perception is **contrast**, which in is the difference in appearance of two or more parts of a visual field seen simultaneously, or successively. As such contrast can exist in such visual parameters as brightness, lightness, and color. Contrast always contains some type of visual information, making it an essential performance feature of any electronic visual display. Factors influencing the contrast of electronic visual displays include the electrical input driving signal (analog or digital), the viewing direction, and the ambient illumination. Electronic visual displays exhibit several forms of contrast, including:

- **Luminance contrast** — This is the ratio between the higher luminance (LH), and the lower luminance (LL), which defines the detected feature. This ratio is often improperly called the contrast ratio (CR), when used to specify the contrast of electronic visual display devices featuring high luminances. However, this parameter is actually a luminance ratio. The luminance (contrast) ratio CR is a dimensionless number, and is usually indicated by adding ":1" to the value of the quotient, such as a contrast ratio of 900:1. As such, a singular contrast ratio (CR = 1) would signify no contrast at all.

- **Color contrast** — Although two parts of a visual field can be of equal luminance, their chromatic composition (color) will be different. A suitable chromaticity system will describe such a color contrast by a distance. With electronic displays, the term used to denote color contrast is the color difference.

- **Full-screen contrast** — This involves time sequential measurements of the luminance values used for contrast evaluation. The active area of the display screen is often completely set to one of the contrasting optical states for which the measurement is to be determined. For example, a. completely white screen will be set as R=G=B=100%, while a completely black screen will be set to R=G=B=0%. Luminance is measured from one screen to the next, time sequentially. This measurement assumes the display device is not exhibiting loading effects, such as found in CRT displays and Plasma Display Panels (PDPs).

Full-swing contrast

- **Full-swing contrast** − Any two non-identical test patterns are suitable for evaluating the contrast between them. If one test pattern comprises a completely bright state of R=G=B=100%, and the other uses a completely dark state (full-black) of R=G=B=0%, the resulting full-swing contrast is the highest (maximum) contrast the display can achieve. Full-swing contrast is assumed when no test pattern is specified in a data sheet contrast statement.

Static contrast

- **Static contrast** − For static contrast evaluation, first apply a test pattern to the display's electrical interface and wait for the optical response to settle into a stable steady state. Then measure and record the luminance of the first test pattern. Apply the second test pattern to the display's electrical interface and wait until the optical response has settled to a stable steady state. Measure and record the luminance of the second test pattern. Then, calculate the resulting static contrast for the two test patterns using either the CR, CM, or K metric.

Transient contrast

- **Transient contrast** − If luminance is measured before the optical response has settled to a stable steady state, the resulting measurement is that of a transient contrast, rather than a static one. This is often the situation when the image content is rapidly changing. During the display of video or movie content, the display's optical state may not reach a stable steady state because of slow response. Compared to the static contrast, the contrast of an unsettled optical response is reduced.

Dynamic contrast

- **Dynamic contrast** − This is a technique for improving or expanding the sequential contrast of LCD-screens by backlight modulation.

Dark-room contrast

- **Dark-room contrast** − In order to measure the highest contrast possible, the dark state of the display under test must not be corrupted by light from its surroundings, because even small increments of ambient illumination will reduce the measurement considerably.

Ambient contrast

- **Ambient contrast** − This is contrast that can be experienced, or measured, in the presence of ambient illumination. Keep in mind that a special type of outdoor ambient contrast exists when the illumination is very intense, up to 100,000 lux, called "daylight contrast". The fact is that dark areas of a display are always corrupted by reflected light. Therefore, the maintaining of reasonable ambient contrast values can only be achieved when the display is provided with anti reflection and/or anti-glare coatings.

Concurrent contrast

- **Concurrent contrast** − When a test pattern is displayed that simultaneously contains dark and light areas, such as a checkerboard pattern, the apparent contrast is called concurrent. Contrast values obtained from two subsequently displayed full-screen patterns may differ from those taken from a checkerboard pattern using identical optical states, due to non-ideal properties of the display-screen, such as crosstalk, or some type of problem with the light measuring device.

Successive contrast

- **Successive contrast** − When a contrast is established between two optical states that are perceived or measured one after the other, this is called successive contrast. An example of successive contrast would be those measurements taken between two full-screen patterns.

CONTENT MANAGEMENT SYSTEMS

Content Management System

A **Content Management System** is a computer software system for organizing and managing information. There are several types designed for storage and managing of documentations, technical manuals, sales guides and other specific web applications.

Content management systems are also used to provide a central computer-based system for storage and management of audio and video files for home entertainment systems. This concept is illustrated in Figure 1-39. The following is a listing of the types of content management systems you will need to know about. They include:

- Media servers
- Media PC or Home Theater PC
- MP3 players
- DVD players
- Satellite
- DVR
- Satellite radio
- Streaming media

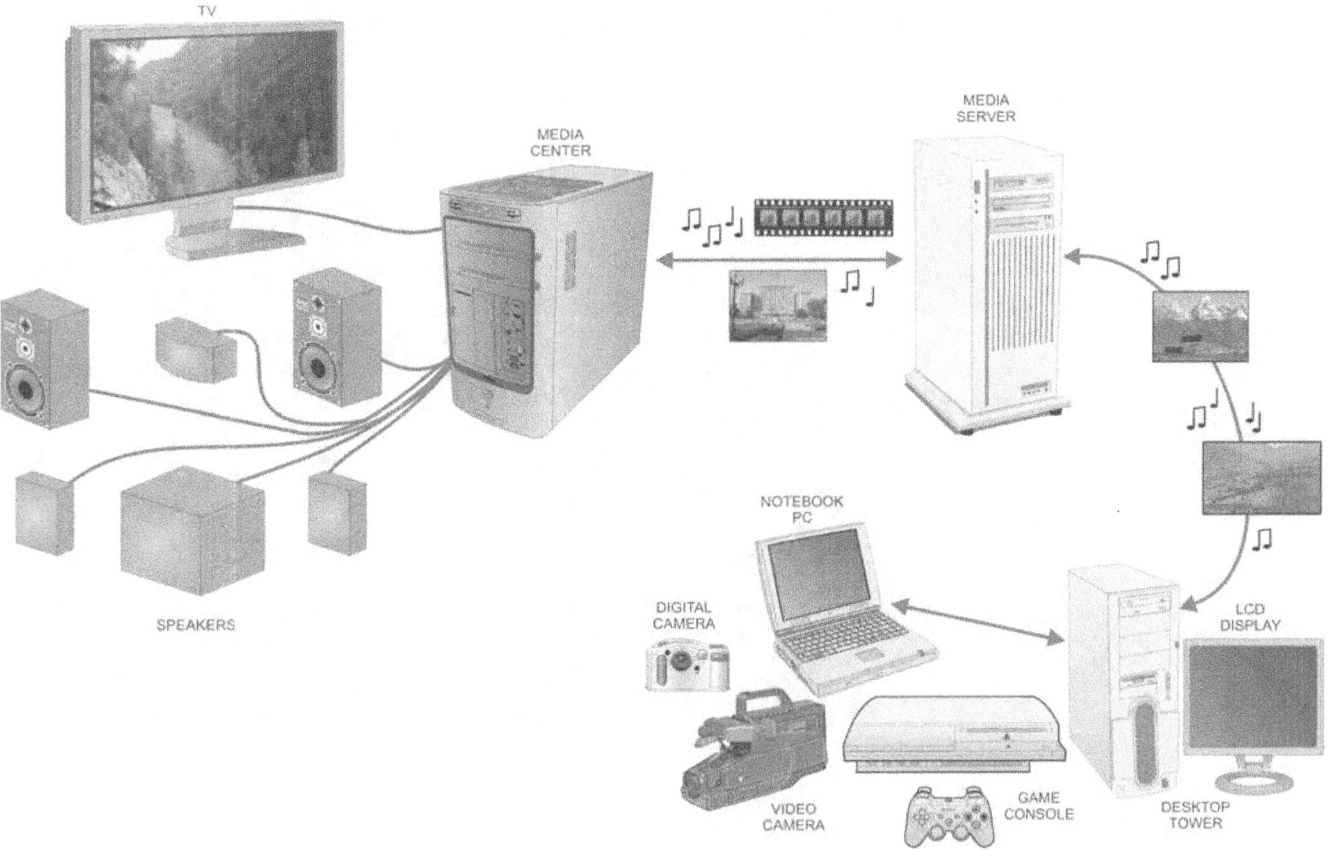

Figure 1-39: Content Management Devices

Media Servers and Extenders

A residential **media server** computer is installed in the residence as the center of the home communications system. Its main responsibility is to store all of the home's digital media—its music, recorded TV, family pictures, home movies, etc. and deliver them to designated devices in the house for display or presentation.

Media extender technology is built into various home entertainment devices, such as TVs, DVD players, and various quiet components. This technology permits the client to install his or her PC in its most sensible location, and use it as a digital entertainment hub to network various media components throughout the residence. For the Windows Media Center, extender devices include set-top boxes configured to connect over a network link to a computer running a recent version of Windows Media Center. The computer's media center functions are streamed to the extender device, permitting someone to view photos, play videos, listen to music, watch live television using DVR functions, and watch recorded TV programming from a remote television, or on some other display device.

Home Theater Personal Computer

A **home theater personal computer** (**HTPC**) or media center PC, as illustrated in Figure 1-40, is a convergence device that combines the functions of a personal computer and a digital video recorder (DVR). It is connected to a television or a television-sized computer display and is often used as a digital photo, music, video player, TV receiver and digital video recorder. Home theater PCs are also referred to as media center systems or media servers.

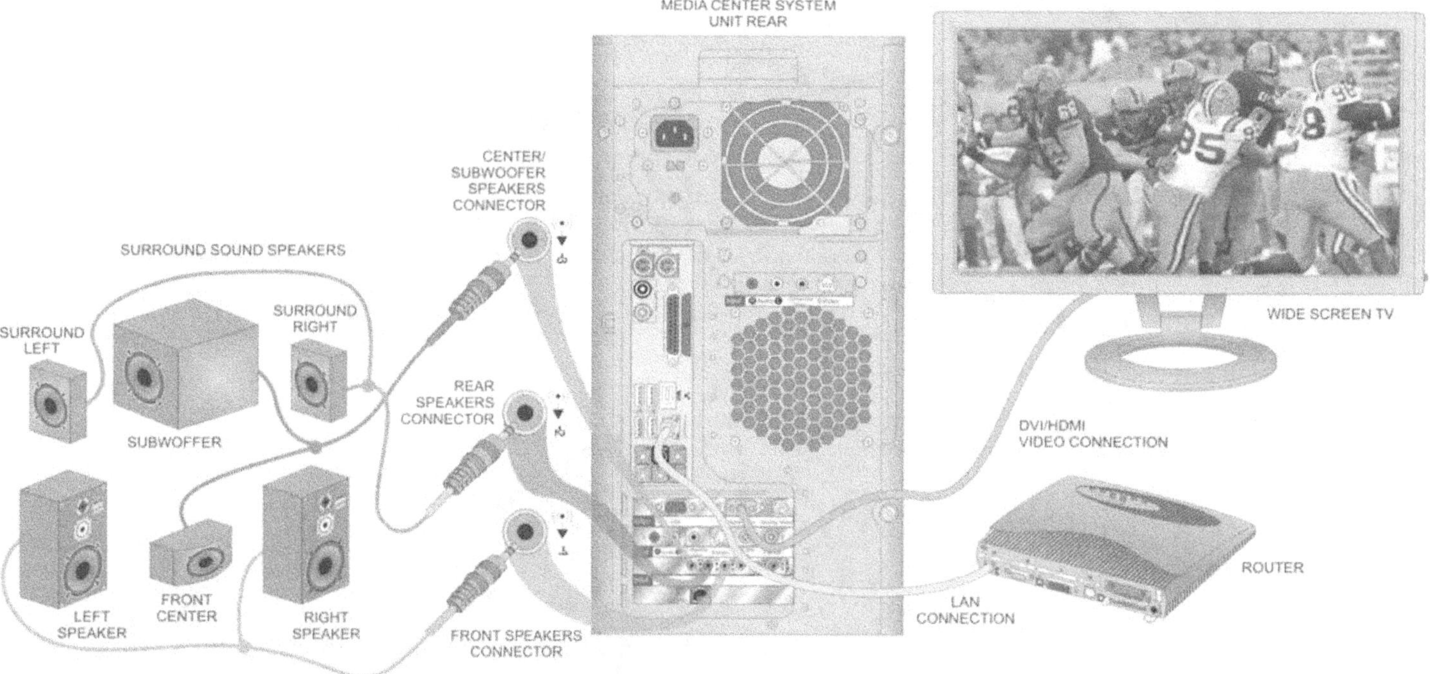

Figure 1-40:
Media Center PC

Home Theater PC Operating Systems

A common approach for Windows based Media PCs is to install a version of Windows Vista that contains the Windows Media Center (Windows Vista Home Premium or Windows Vista Ultimate) or Windows XP Media Center Edition as the operating system. The Windows Media Center Start page is depicted in Figure 1-41.

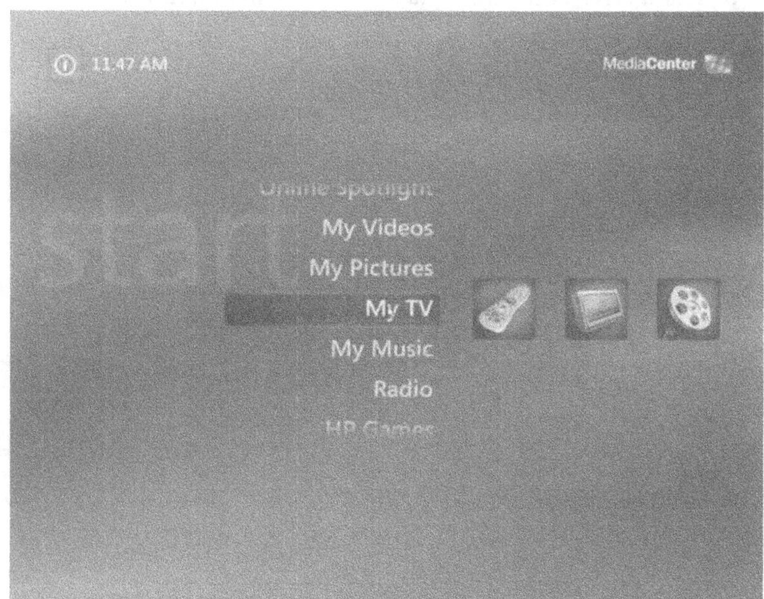

**Figure 1-41:
Media Center Start
Page Options**

MP3 Players

MP3 players are used for the storage and playback of audio files that have been encoded using the **Motion Picture Experts Group** (**MPEG**)-1 Audio Layer 3, more commonly referred to as MP3. This encoding format is used to create an MP3 file, a way to store a single segment of audio, commonly a song, so that it can be organized or easily transferred between computers and other devices such as MP3 players.

MP3 uses a lossy compression algorithm that is designed to greatly reduce the amount of data required to represent audio recordings, yet still sound like faithful reproductions of the original uncompressed audio to most listeners. An MP3 digital file created using the mid-range bit rate setting of 128 kbps results in a file that is typically about 1/10th the size of the compact disc file created from the same audio source.

MP3 players are pocket sized devices that support storage and playback of music and other sound files. While there are several types of these players, they all can be placed into one of three categories: hard drive based players, micro hard drive based players and flash based players. A legacy type player is the standard portable CD player that has been replaced by smaller MP3 players

Hard Drive-Based MP3 Players

Hard drive-based MP3 players often utilize 1.8-inch (or 2.5-inch) hard drives, and come with a rather large color LCD screen as well as image and video playback capabilities. They are available with 10 to 100 GB of storage.

An example of a hard drive-based MP3 player is shown in Figure 1-42.

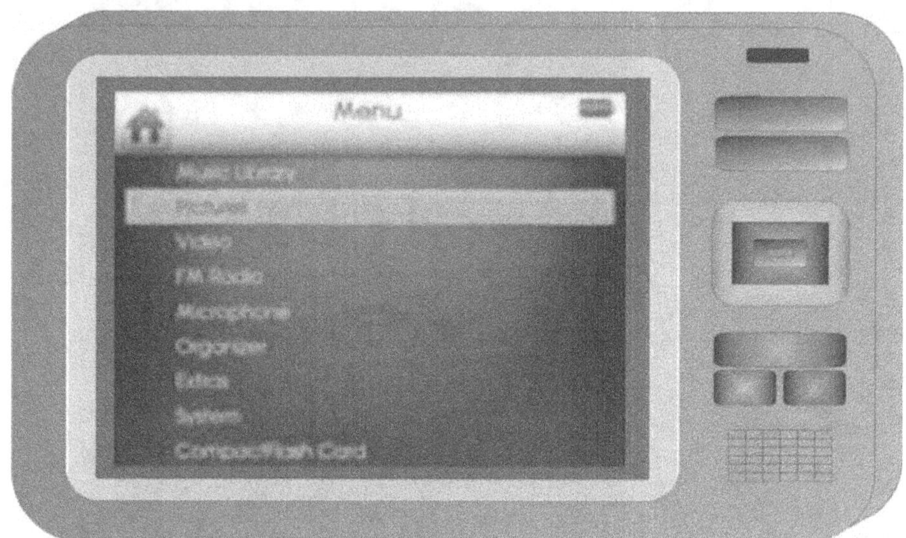

Figure 1-42: Hard Drive-Based MP3 Player

Flash Memory-Based MP3 Players

The capacity of contemporary flash memory-based MP3 players varies from 256MB (or lower) to 8GB (or higher). Since there are no mechanically moving parts in this type of MP3 player shock and vibration protection becomes less important. The power consumption of flash memory is also lower than that of hard drives, thereby making possible longer playback times.

There are also certain MP3 player products that feature no built-in memory/storage, and require the user to insert their own flash memory card (SD/MMC card most often) or flash drive to operate. A flash memory-based MP3 player is shown in Figure 1-43.

Figure 1-43: Flash Memory-Based MP3 Player

CD-Based Players

Portable CD players that can decode and play MP3 audio files stored on CDs are typically called "CD players" as opposed to MP3 players.

A CD-based MP3-capable player shown in Figure 1-44 is essentially a common portable CD player with MP3 playback support (may also support other digital audio formats). They are not as small and portable as the standard MP3 players.

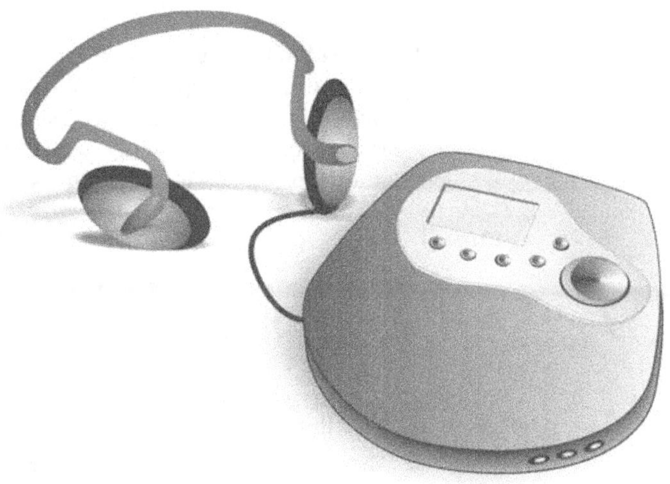

Figure 1-44: CD-based Player

DVD Players and Recorders

Digital Versatile Disk (DVD) player/recorder shown in Figure 1-45 provide high-quality video in a home theater system. The precise definition of DVD has been modified by the industry from its original title of digital versatile disk to a more marketable term; digital video disk.

A DVD player decodes the digital information stored on the disk and converts it to analog video and audio information for processing by an NTSC standard television receiver. DVDs. DVD players can also play audio CDs.

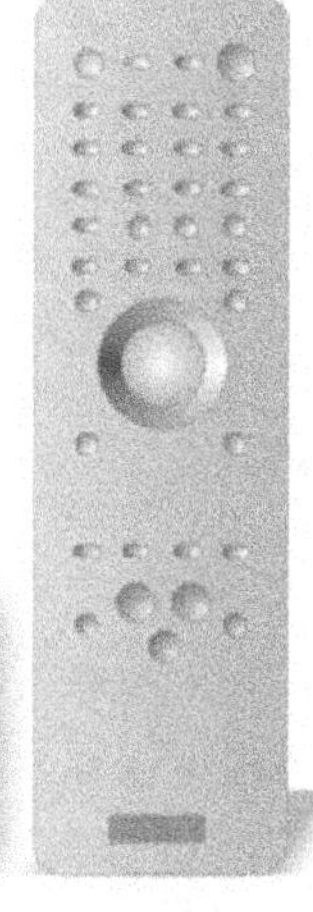

Figure 1-45: DVD Player /Recorder

DVD recording technology uses a red laser to permanently transform a dye-recording layer on the DVD-R disc, similar to how CD-R recording works. This permanent transformation of the media is characteristic of a write-once (DVD-R) format.

The rewritable format is the DVD-RW variant, formally known as DVD Re-Recordable. Like the DVD-Video format, the video information is written as a single track, starting from the innermost portion of the disc and spiraling out to the outer edge.

Progressive or noninterlaced scanning is any method for displaying, storing or transmitting moving images in which all the lines of each frame are drawn in sequence. This is in contrast to the interlacing used in traditional television systems where only the odd lines, then the even lines of each frame are drawn alternatively (each image now called a field) are drawn.

Before HDTVs became common, players were sold which could output 480 scan lines in progressive format (480p). TVs with this feature were often in the upper price range of a manufacturer's line. To utilize this feature, a TV or other display with a progressive scan input was needed. HDTVs usually have a progressive scan input; progressive scan inputs are less common on standard definition TVs.

Blu-ray Disc

Blu-ray Disc (also known as **Blu-ray** or **BD**) is a high-density optical disc format for the storage of digital information, including high-definition video (HDTV). The name Blu-ray Disc is derived from the blue-violet laser used to read and write this type of disc. Because of its shorter wavelength (405 nm), substantially more data can be stored on a Blu-ray Disc than on the DVD format, which uses a red (650 nm) laser. A Blu-ray Disc can store 50 GB, almost six times the capacity of a DVD. Figure 1-46 shows how different types of laser light affect data storage on a disc (shorter wave length light produces higher densities).

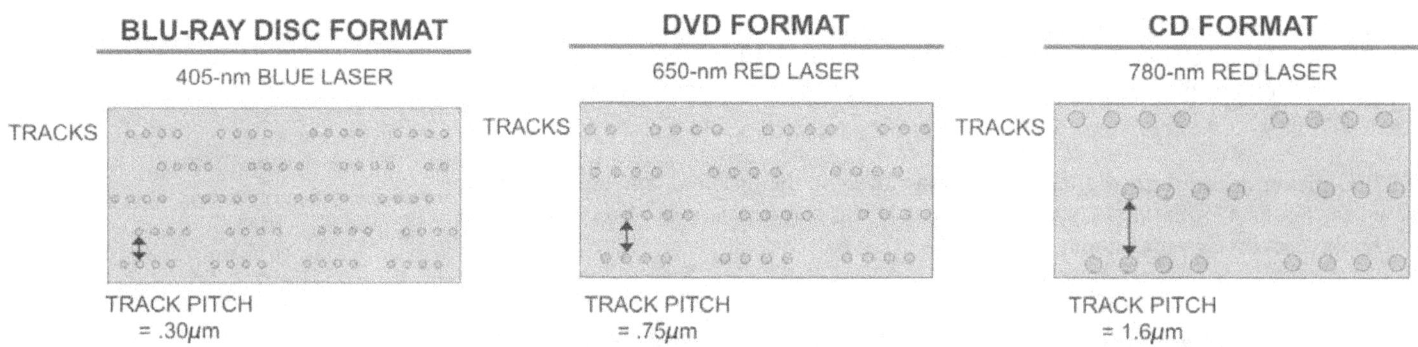

Figure 1-46: Blue vs. Red Lasers

DVD Player Video Connectors

DVD players have standard RCA composite video, S-video, and component video outputs. On most DVD players, the component video outputs can be set to transfer either a standard interlaced video signal or a progressive scan video signal to a television. Some DVD players also have **Digital Visual Interface (DVI)** or **High Definition Multimedia Interface (HDMI)** outputs for better connection to a HDTV as well. Figure 1-47 shows typical connections for a DVD player.

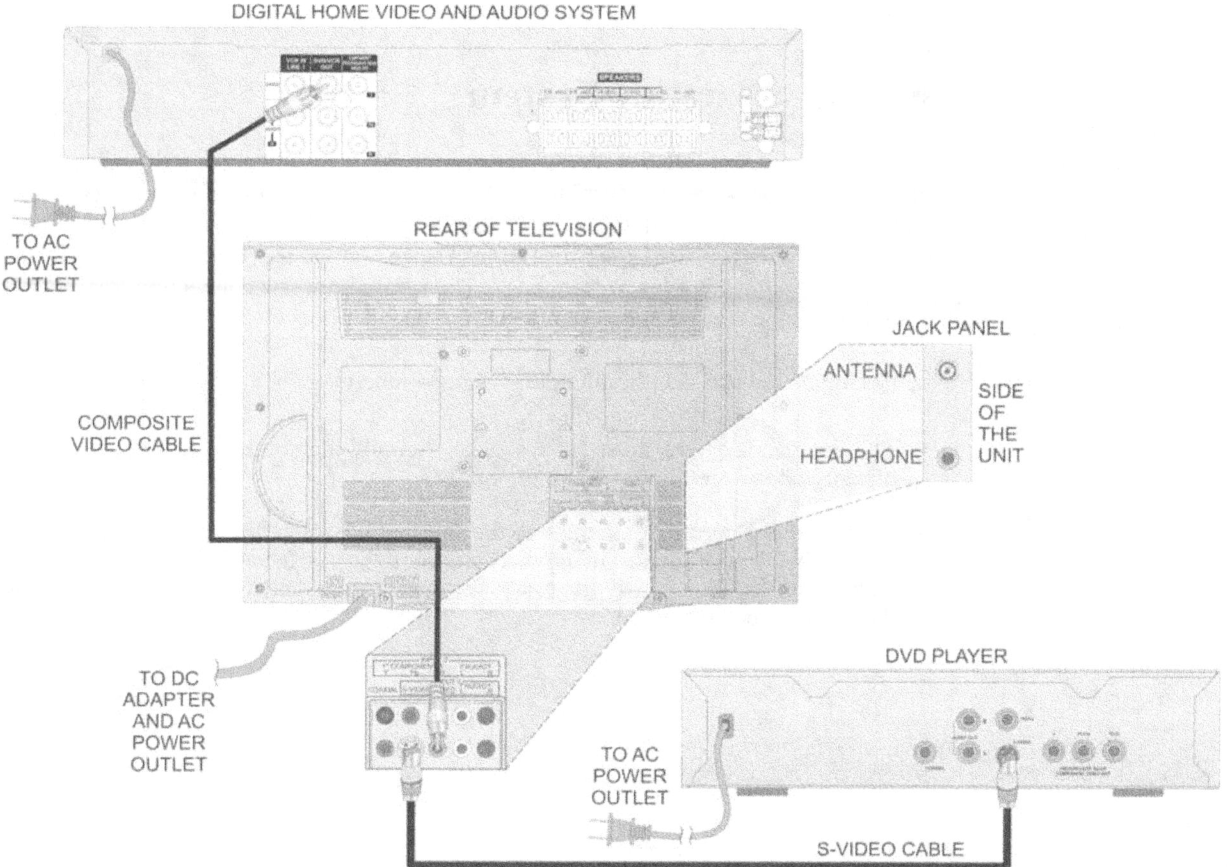

Figure 1-47: Video Connection Diagram

Satellite Receivers

Satellite receiver/decoders, like the one depicted in Figure 1-48, provide the digital decoding and unscrambling of the downlink from a **direct broadcast satellite (DBS)**. They connect the antenna with the A/V receiver, which provides the video and Dolby Digital sound to the home theater system. DBS satellites provide an alternative source for subscription TV service for home audio/video systems.

direct broadcast satellite (DBS)

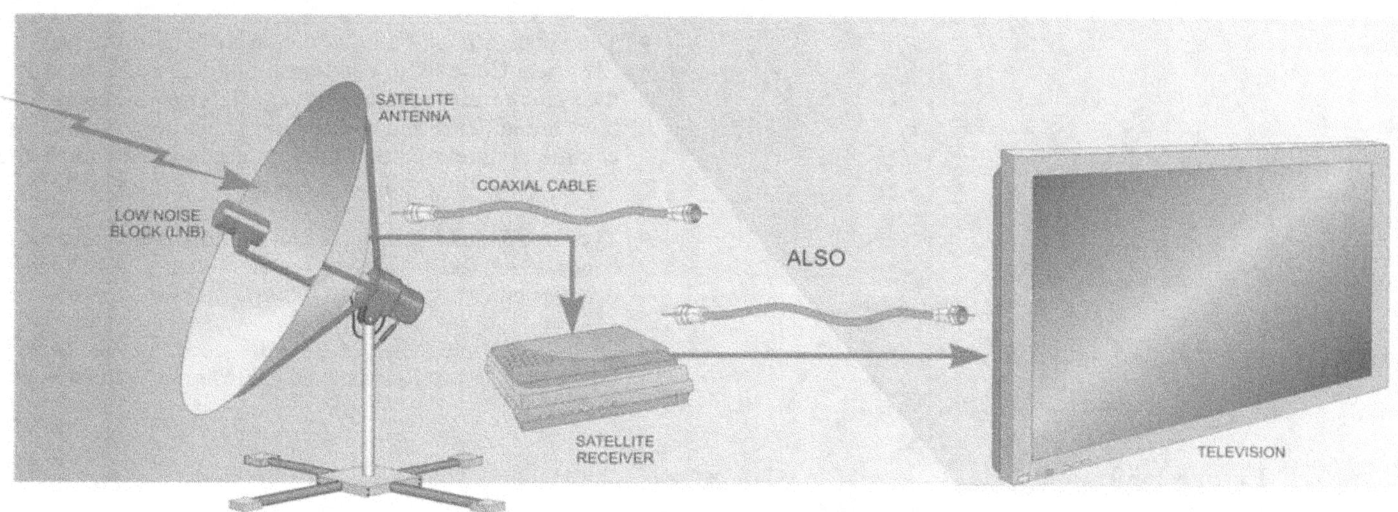

Figure 1-48: Satellite Receiver/Decoders

Installing a Satellite System

Direct broadcast satellite (DBS) outdoor antennas must be installed using the instructions included with the hardware kit. Most DBS antennas include provisions for making the final adjustments using the signal strength meter display on the TV receiver screen. DBS dish antennas are equipped with an **Low Noise Block Down Converter (LNB)** that converts the received satellite signal to a lower frequency. The LNB requires a connection to a coaxial cable for sending the received signal to the indoor decoder box.

Guidelines for installing a satellite antenna system for receiving DBS service are as follows:

- The antenna location must have a clear view of the southern sky. The antenna distance above the ground is not important.

- Determine the correct look angle to the selected DBS satellite from the installation geographical coordinates using the program that is usually included with the installation kit. Figure 1-49 illustrates the two coordinates of azimuth and elevation that must be calculated.

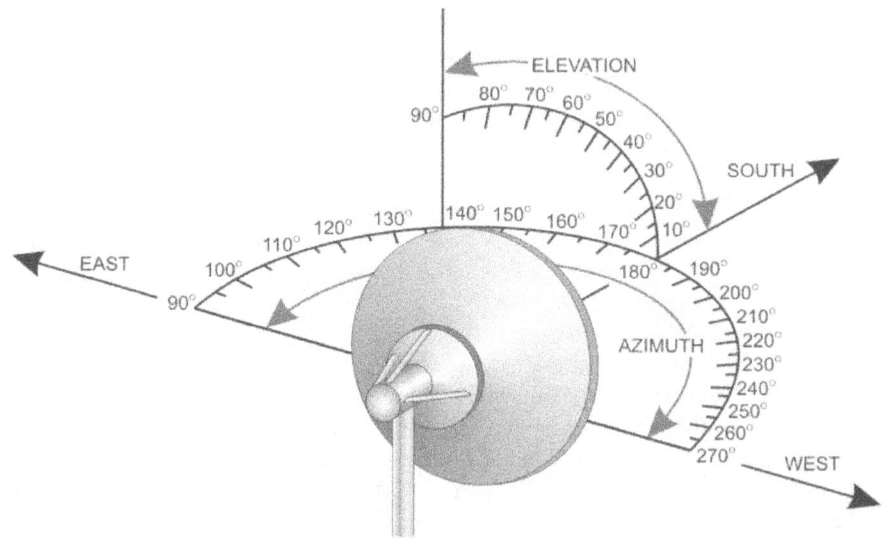

**Figure 1-49:
Antenna Azimuth
and Elevation**

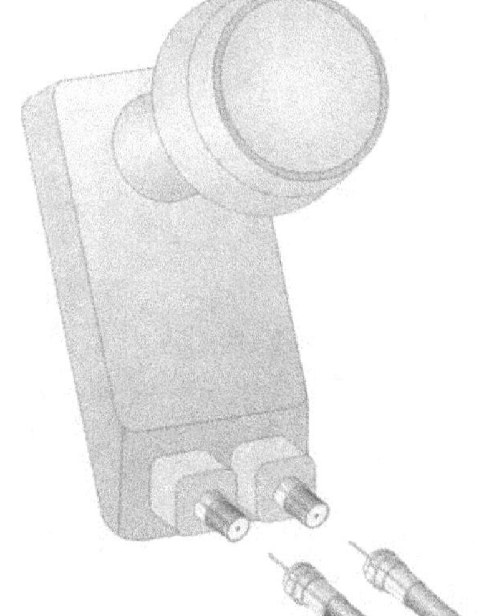

**Figure 1-50:
Dual LNB**

- The antenna must be grounded according to the National Electrical Code (NEC) and local building codes to protect the system from lightning damage. It is very important that both the antenna and the coaxial cable be grounded to a central common-point building ground with an 8-foot copper grounding rod.

- Type RG6 quad-shield coaxial cable should be used to connect the feed assembly on the antenna to the indoor decoder box.

- Dual LNB antennas designed for feeding two decoder boxes require two RG6 coaxial cable feeds, as illustrated in Figure 1-50.

- The multisatellite triple LNB antenna, as shown in Figure 1-51, is capable of receiving signals from three satellite locations (at 101°, 110°, and 119° west longitude) simultaneously. The dish comes included with an integrated 4x4 multi-switch for easy connection to four receivers.

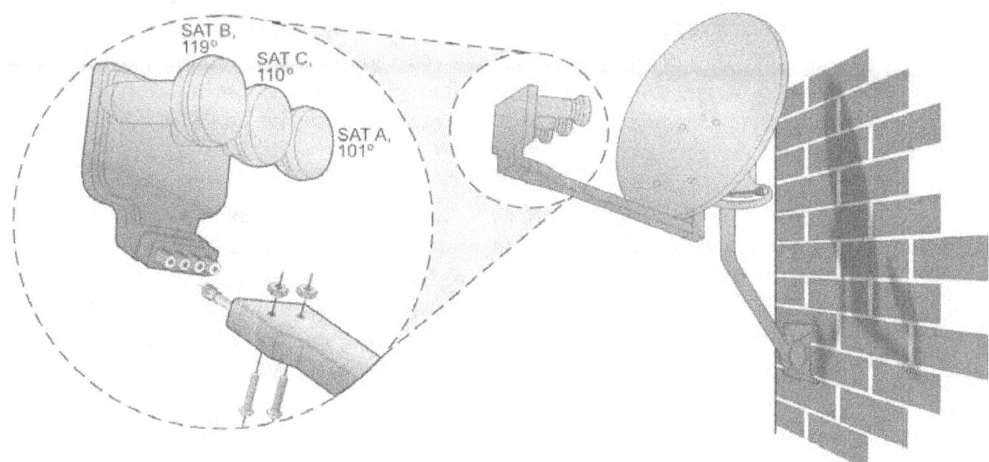

Figure 1-51: Triple LNB Satellite Antenna

> **NOTE**
>
> The term **LNB** is an acronym for **low noise block-down converter** (so called because it converts a whole band, or block, of frequencies to a lower band). A triple LNB with an elliptical dish can receive and down-convert signals for all DBS satellites located in the orbital slots at 101°, 110°, and 119° west longitude without moving the antenna.

Diplexer

A **diplexer** is used outside to combine two separate signals: one signal from the satellite dish and another signal from either an off-air TV antenna or cable TV feed into a single coaxial cable. A second diplexer is used inside near the satellite receiver to separate the combined signals. **Satellite diplexers** eliminate the need for an extra coaxial cable transmission line.

CHALLENGE #1

You are selecting the components and installing a DBS satellite antenna and decoder for a friend. He wishes to connect two separate receiver decoder boxes for the home theater and bedroom areas. Describe the equipment, cabling, and installation planning needed to meet these requirements.

Personal Video Recorders

Personal Video Recorders (**PVRs**) are similar to Video Cassette Recorders (VCRs) except that they use a hard disk instead of a magnetic tape as the recording media. They are designed to record live TV broadcast streams "on the fly" and allow the viewer to pause a program, take a break, and resume watching a program even though the broadcast program has continued past that point. A viewer may receive a phone call without missing a favorite play during a football game. It is possible with a PVR to put a favorite live cable broadcast sports event on hold, similar to the pause function on a VCR.

---TEST TIP---

Know that a personal video recorder stores streaming video on a hard disk and that TiVo and Replay TV are the scheduling and programming service providers.

Basically, the PVR consists of a TV tuner, a fast hard-disk drive, a proprietary computer board (with a fast processor and lots of RAM) and video encoder/decoder modules, as illustrated in Figure 1-52. PVRs use compression algorithms to make streaming video recording feasible. In addition, PVRs provide more versatile recording options than do VCRs. They can record, pause and resume from anywhere in the program stream similar to a DVD.

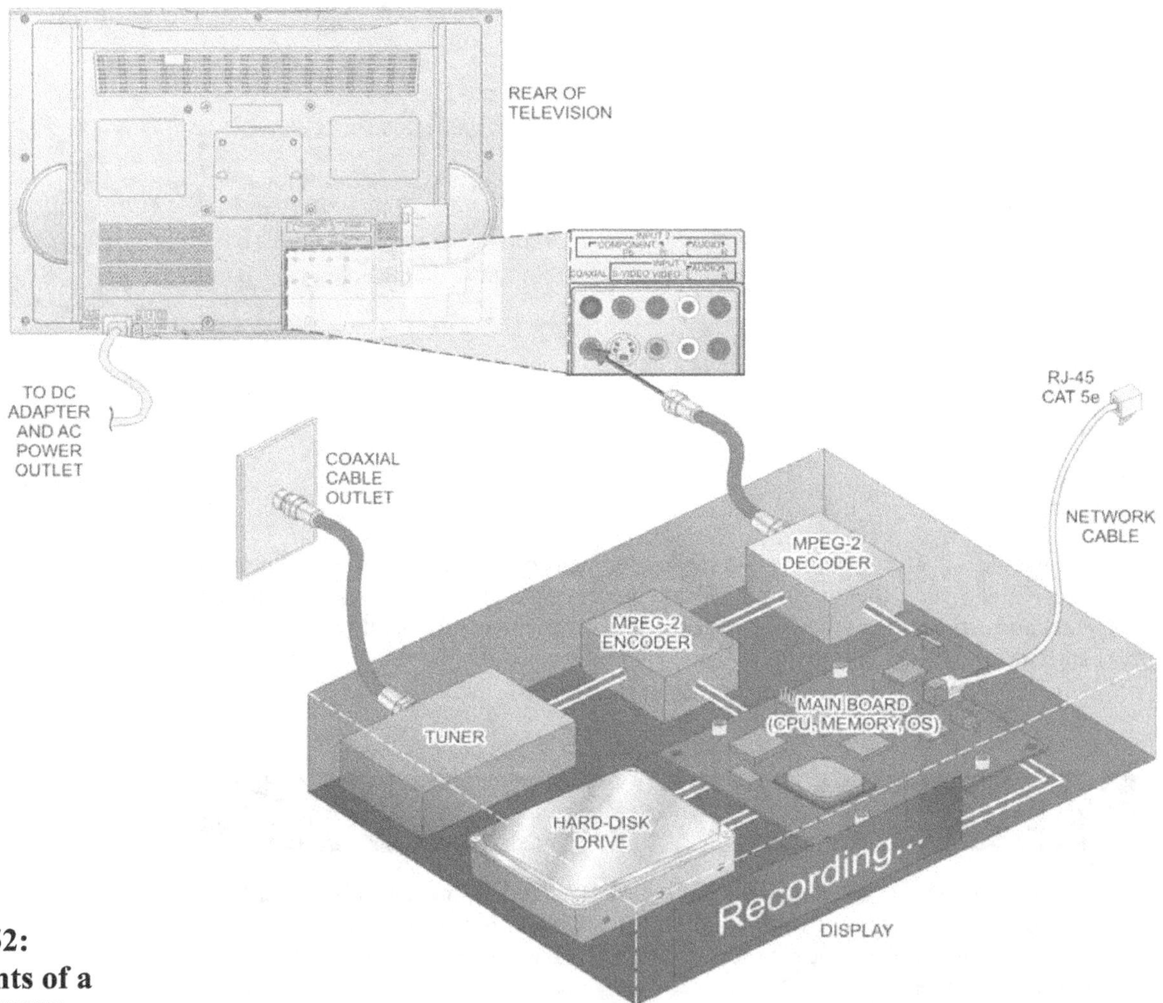

**Figure 1-52:
Components of a
PC-based PVR**

PVRs require the user to obtain subscriber service for scheduling and programming service. TiVo and Replay TV are competing PVR service providers.

Satellite Radio

Satellite radio is a digital radio service that is broadcast by a communications satellite, which covers a much wider geographical area than terrestrial radio. Providers usually carry a variety of news, weather, sports, and music channels, with the music channels generally being commercial-free.

Each satellite radio receiver has an Electronic Serial Number (ESN) Radio ID to identify it. When a unit is activated with a subscription, an authorization code is sent in the digital stream telling the receiver to allow access to the blocked channels.

Streaming Audio and Video

Various Web sites provide program content from radio and television broadcast stations that can be received by a home computer connected to the Internet. The term "streaming" refers to the audio/video files' ability to play as they are being received by a home computer, so that the home user does not have to wait for a file to download. **Streaming audio and video** files are compressed so they require less storage space and bandwidth than non-streaming uncompressed files. As shown in Figure 1-53, streaming audio/video files can be processed in real time. When audio or video is streamed, a small buffer space is created on the user's computer, and data starts downloading into it.

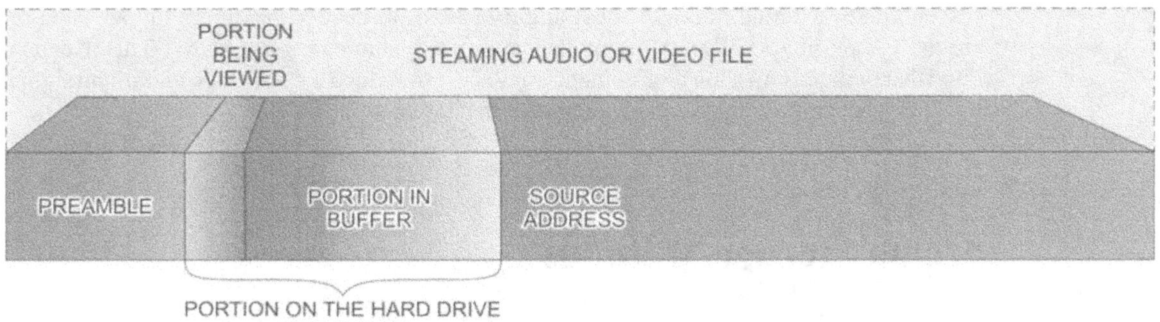

As soon as the buffer is full (usually just a matter of seconds), the file starts to play. As the file plays, it uses up information in the buffer, but while it is playing, more data is being downloaded. As long as the data can be downloaded as fast as it is used up in playback, the file will play smoothly. Streaming files require an application program on a computer to perform the processing on the received files. Several applications are available including QuickTime, RealPlayer, and Windows Media Player. Streaming audio and video is made possible due to the development of very efficient compression standards that reduce the number of bits required to produce good-quality video and audio at modest transmission bit rates.

Figure 1-53: Streaming Audio/Video File Processing

While modern Ethernet systems utilize UTP cable for their networks, keep in mind that the original 10 Mbps Ethernet used Manchester encoding. Fast Ethernet (at 100 Mbps) used a 4b/5b Multi-Level Transmit (MLT) code. Gigabit Ethernet utilizes 8b/10b encoding for more efficient use of the limited cable bandwidth, sending 1 Gbps within the approximately 100 MHz of bandwidth capacity stated for a UTP CAT5e cable.

Compared to CAT5/5e UTP cabling, the newer CAT6 variety represents a significant improvement for networking, providing capabilities that exceed its specified frequency range of 250 MHz in practice. Its residential applications include cable/satellite TV, stored or live video broadcast, distance learning, security monitoring, and video conferencing.

Broadband video (CATV or cable television) carries a range of signals extending beyond 650 MHz, with RG-59 or RG-6 coax commonly used for home networks carrying these frequencies. Not widely understood is the fact that high performance twisted pair cabling can also support broadband video streaming. It is capable of complying with the stringent requirements of CATV regarding signal levels, alien noise, and signal distribution. Compliance with regulatory emission standards include signal level measurements, noise ingress measurements, and tilt adjustments for video and audio carrier levels across the frequency band, up to 550 MHz. Broadband video (CATV or cable television) carries a range of signals extending beyond 650 MHz, with RG-59 or RG-6 "bare copper (BC)" coax commonly used for home networks carrying these frequencies.

Broadband video over twisted pair is capable of transmitting multiple channels simultaneously, using only a single pair in a 4-pair cable for broadcast video distribution, or two pairs for bi-directional interactive video transmission. Current broadband video systems use unbalanced 75-ohm coaxial cables, which requires a high quality broadband balun to transmit these signals over 100-ohm balanced twisted pair cabling. The balun adapts a 75-ohm coaxial input (F-connector) to a 100-ohm balanced output using a modular 8-position connector.

The main concern for residential clients is how far the video signals can be streamed over CAT5e/6 cabling. Although most TV receivers accommodate a wide dynamic range of signals, the minimum acceptable signal level at the remote TV is near 1 mV pk-pk (–10 dBmV). A signal weaker than this will produce a snowy picture that is much more susceptible to external noise. A local amplifier will produce a maximum output level of around 1 V pk-pk (50 dBmV), giving the cabling a dynamic range of up to 60 dB. Keep in mind that radiated emission requirements may serve to reduce the 50dBmV maximum signal level, further limiting the dynamic range for the application.

Media Storage Methods

Two popular methods of storage are memory cards and network attached storage.

- A memory card or flash memory card is a solid-state electronic flash memory data storage device used with digital cameras, handheld computers, telephones, music players, video game consoles, and other electronics. They offer high re-record-ability, power-free storage, small form factor, and rugged environmental specifications. There are also non-solid-state memory cards that do not use flash memory, and there are different types of flash memory.

- Network Attached Storage (NAS) is mass storage attached to a computer which another computer can access at file level over a local-area network, a private wide-area network, or in the case of online file storage, over the Internet.

Although computer-based information is easier to store and share than legacy paper-based systems, clients can risk getting lulled into a false sense of security. Remember that computer drives cannot offer data permanence. This makes it important that crucial data be backed up regularly. Data loss invokes a feeling similar in many respects to that of being robbed. The sense of loss can be devastating for either a small business owner, or the home technology enthusiast with an extensive A/V collection. Important emails and family photos are just as irreplaceable to a home technology enthusiast as customer information files are to a small business owner. In order to protect valuable personal or business data, it is vital to utilize a reliable system for saving, backing up, and archiving. Anticipating the possibility of periodic disaster helps to ensure that data is backed up, rather than being lost forever. Options for saving and storing your data include:

- *Backing up on CDs or DVDs* — Although CDs and DVDs have considerably more capacity than floppies used to, the size of modern hard drives will require the burning of many CDs or DVDs. Locating and restoring files will also be time-consuming, regardless of the filing system used. Keep in mind that storing backup archives in the same location as the computer hard drives may result in the destruction of both at the same time.

- *Backing up to flash memory* — Memory cards and flash drives provide more space and quicker backup times than CDs and DVDs. Their portability enables copies of the data to be taken to alternate locations for total data security. However, memory cards and flash drives often do not offer as much storage capacity as might be required.

- *Backing up to an external hard drive* — This is a preferred solution due to the fact that external hard drives often have as much (or more) storage space than the original hard drive. Rapid data transfer times are realized through USB connections, while separate storage locations will ensure against simultaneous destruction of original and backup files.

- *Using an online backup or archiving service* — This is often the easiest and safest data backup method, making it worth the expense to many people, consisting of low monthly storage fees as opposed to the purchase of storage devices. Services include the setting of automatic backups, with the added assurance that the data is stored at some distance from the home.

AUDIO/VIDEO SYSTEM FORMATS AND CONNECTIONS

Video connections on the rear panel of video consumer electronic devices are configured with different jacks for connecting AM/FM receivers, television receivers, DVDs, VCRs, PC monitors, laserdisc players, and camcorders. This section describes the various formats for both audio and video connections.

Video formats and connector types include:

- Component video

- Composite video

- S-Video

- RGB video

Audio formats and connector types include:

- Analog audio
- Digital audio

Other audio/video connectors include:

- RCA connectors
- Binding posts
- F-type connectors

The detailed descriptions of these formats and audio/video connectors are given in the following paragraphs.

Composite Video

Composite video is a video format in which color (chrominance), luminance (brightness), and synchronization signals are mixed into a single radio carrier wave.

It is the video standard used for television broadcasting. Inside the television receiver, the composite signal is separated into the red, green, and blue (RGB) color information. Composite video is the most widely used analog standard for connecting TVs with VCRs, laserdisc players, and camcorders. As illustrated in Figure 1-54, a single coaxial cable with an RCA plug is required to connect audio/video equipment using the (yellow color coded) composite video jack.

Figure 1-54: Composite Video Connection

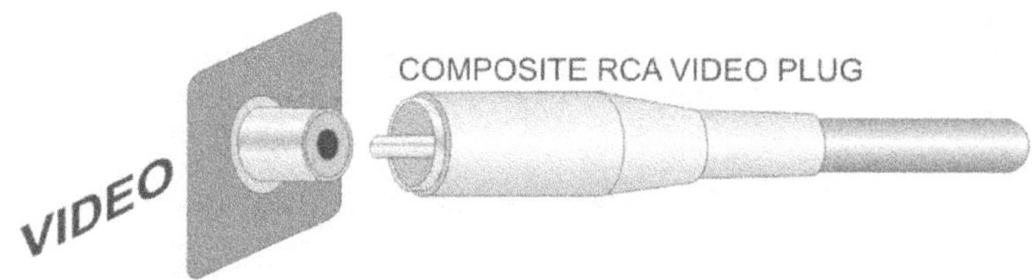

Component Video

Component video is a format that provides the best color reproduction when connecting home theater products. Component video divides the signal into three components called the Y signal and the intermediate color difference signals, R-Y and B-Y.

The luminance signal has a bandwidth greater than 6 MHz and the color difference signals have bandwidths greater than 3 MHz. Component video is used by DVDs to store analog video information in a digital format. It is also the preferred method for connecting DVD players to HDTV receivers.

Component video connections require three bundled 75-ohm coax cables, which carry the Y, R-Y, and B-Y signals separately. Each of the three coax cables is terminated at both ends with RCA connectors.

The component video input jacks to home theater products, shown in Figure 1-55, are color coded with the designations Y (green), Cr (red), and Cb (blue). RGB video is also a component video format, but is used almost exclusively for connecting to video projectors.

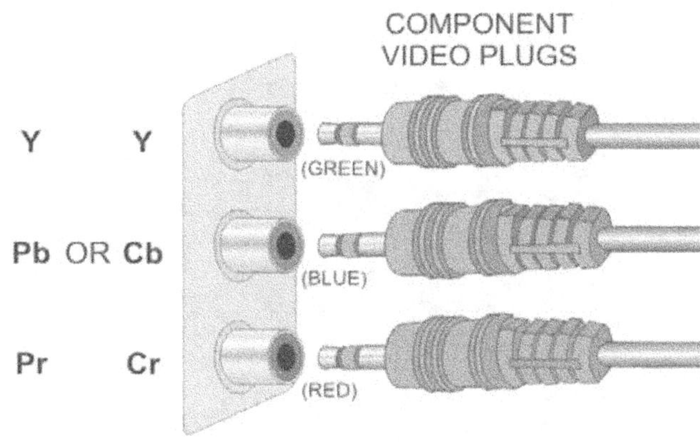

Figure 1-55:
Component Video
Connections

The designations Y/Pb/Pr and Y/Cb/Cr are also used loosely to represent analog component video and may appear on input/output jacks for home theater products.

S-Video

S-Video is a compromise between component analog video and the composite video because it separates the luminance (brightness) and chrominance (color) information. It uses twin coaxial type cables integrated into a single cable and terminated with a DIN connector. DIN connectors are used for video and computer interfaces. DIN is an acronym for *Deutsches Insitut für Normung eV*, a standards-setting organization for Germany.

The S-Video DIN connector configuration is shown in Figure 1-56. Pins 1 and 3 in the figure carry the luminance (brightness) signal, and pins 2 and 4 carry the chrominance (color) information.

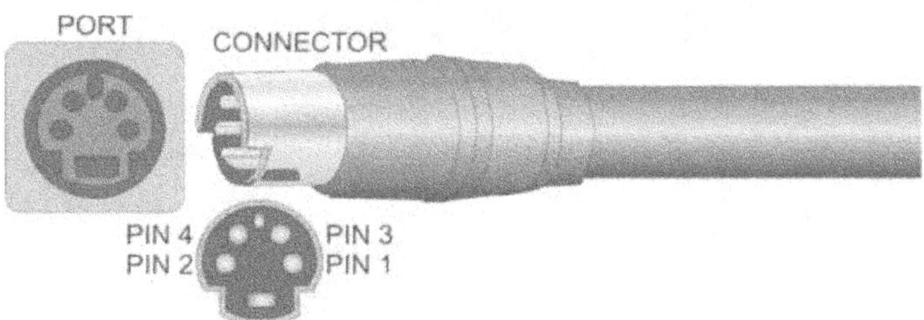

Figure 1-56:
S-Video Connector

RGB

RGB stands for the red, green, and blue signals used to connect computers to monitors and video projectors. There are three different types of RGB cables as follows:

- RGBHV is a five-cable system that splits the video signal for color into red, green, and blue. Two cables carry the sync for the signal (horizontal and vertical sync). Figure 1-57 shows a typical RGBHV connection.

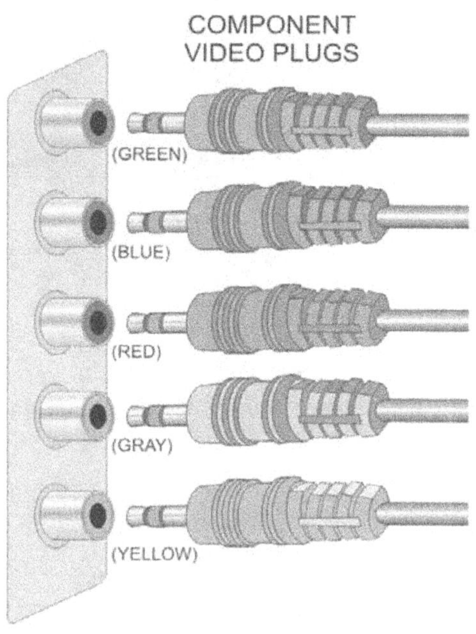

Figure 1-57: RGBHV Connections

- RGBS is a four-cable system that splits the color the same way, but has the horizontal and vertical sync signals on a single fourth cable (black). Figure 1-58 shows a typical RGBS connection.

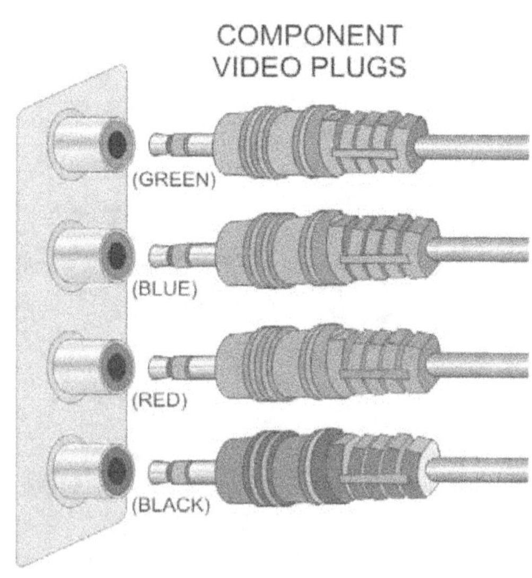

Figure 1-58: RGBS Connection

- Straight RGB video cables again split the color signal into three colors, but carry the additional sync signal on one of the color cables, usually the green (an arrangement called RGB sync on green).

An RGBHV signal is used to connect a computer to a video projector. Five pins on a 15-pin VGA cable are RGBHV signals. The projector recognizes the signal and projects accordingly. A standard 15-pin VGA connector is shown in Figure 1-59.

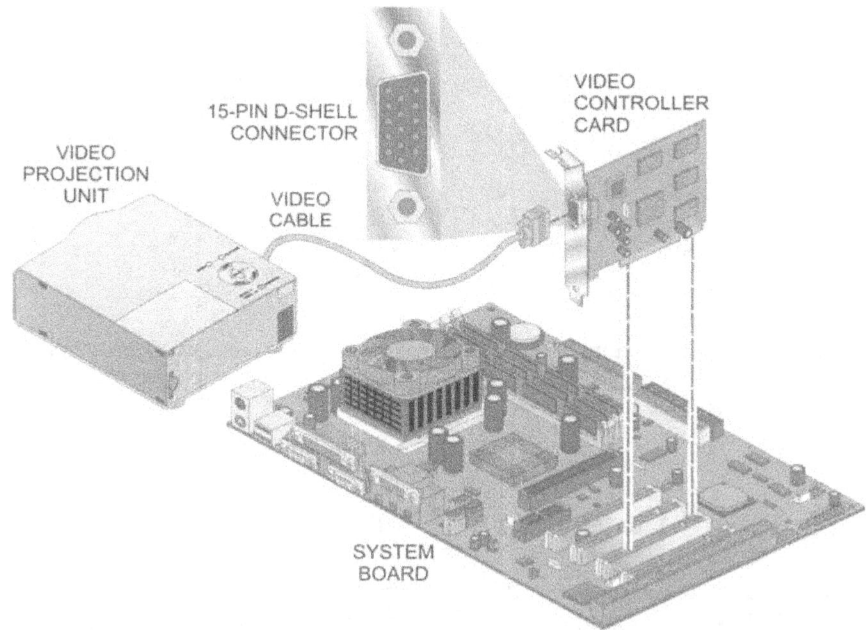

**Figure 1-59:
15-pin VGA
Connector**

Table 1-3 shows the color codes for different RGB connections.

**Table 1-3:
RGB Color Codes**

	RED	GREEN	BLUE	YELLOW	WHITE/GRAY	BLACK
RGBHV	R	G	B	Hsync	Vsync	-
RGB Component (Video Standard)	Y	U	V	-	-	-
	Y	Pb	Pr	-	-	-
	Y	Cb	Cr	-	-	-
RGsB	R	SoG	B	-	-	-
RGBS	R	G	B	-	-	Hsync and Vsync

Analog Audio Component Connections

Analog audio component connections are available on most TVs, VCRs, and CD players. They require an RCA jack with coaxial cable. The analog outputs include left and right stereo or mono, as illustrated in Figure 1-60. Analog connectors are used with audio/video components that are not equipped with the preferred digital audio connectors.

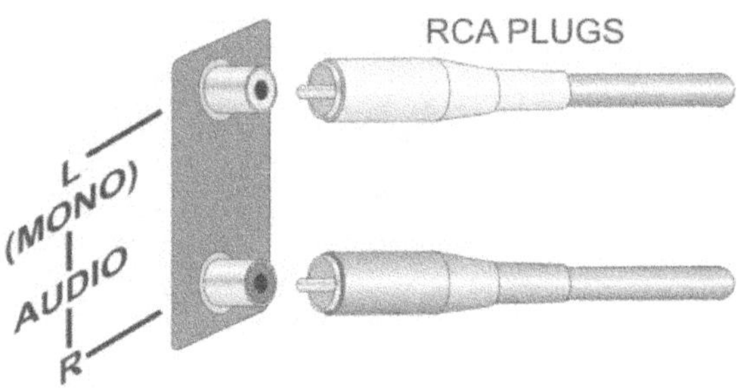

**Figure 1-60:
Analog Audio
Connectors**

Analog outputs available on amplifiers with digital surround sound decoders can be routed to each speaker for surround sound effects in a home theater setting.

Digital Audio Component Connections

Digital audio refers to the transmission or reproduction of sound that is stored in a digital format. CDs, DVDs, Digital Audio Tape (DAT), streaming audio, and sound files stored on a computer are examples of media that contain digital audio and video information. Digital sound files must be processed by a decoder to convert them to analog audio files. Analog audio information can be processed by an amplifier and coupled to a speaker system for the reproduction of sound.

Digital audio output jacks on home theater equipment use either coaxial cable RCA connectors or optical (Toslink) connectors. Toslink is the name of a brand of optical cable developed by the Toshiba Corporation ("Tos" in Toslink stands for the electronics company Toshiba). Toshiba invented the concept of transferring digital audio signals inexpensively and efficiently using clear plastic cable and patented Toslink connectors. CD players, DVD players, mini-disc players/recorders, CD recorders, some MP3 equipment, and some Dolby Digital receivers have Toslink input and output jacks. Given a choice, the fiber optic connection is the preferred method of transferring digital audio signals between devices.

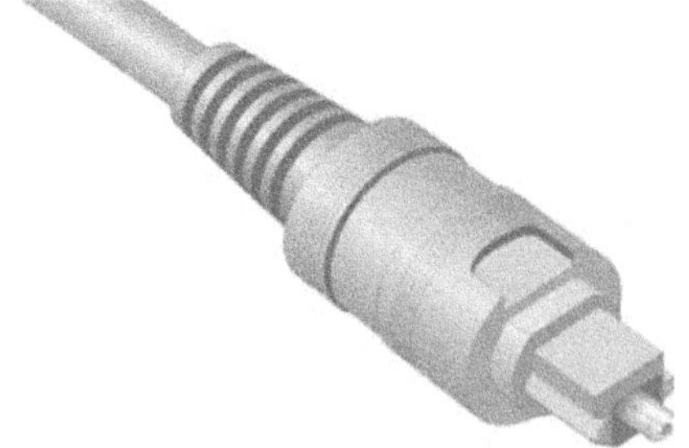

Figure 1-61 shows a Toslink optical cable connector used with most home theater sound system components.

**Figure 1-61:
Toslink Optical Cable**

Audio and Video Cable Terminations

Coaxial cable and UTP cable are used in many home audio and video applications. Table 1-4 summarizes the connectors often used for terminating audio coaxial cables, UTP cable, satellite RG-6 coaxial cable, cable TV coaxial cable, and fiber optic cable.

JACK TYPE	TYPE OF CABLE OR WIRE	APPLICATION
RCA	Coaxial Cable	Analog Audio
RCA	Coaxial Cable	Digital Audio
F-type	RG-6 Coaxial Cable	Satellite/Cable TV
RJ-45	Category 5 UTP	Audio/Digital Data
Binding Posts	Speaker Wire	Audio for Speakers
Toslink	Fiber Cable	Digital Audio

Table 1-4: Audio and Video Cable Terminations and Jacks

High-Definition Multimedia Interface

The **High-Definition Multimedia Interface (HDMI)** is a compact audio/video connector interface for transmitting uncompressed digital streams. It represents the **Digital Rights Management (DRM)** alternative to consumer analog standards such as RF (coaxial cable), composite video, S-Video, component video and VGA, and digital standards such as Digital Visual Interface (DVI). The HDMI connector is shown in Figure 1-62.

Digital Rights Management (DRM) technologies attempt to control use of digital media by preventing access, copying or conversion to other formats by end users. The term is used to describe any technology which makes the unauthorized use of media or devices technically difficult. It does not include forms of copy protection that can be circumvented without modifying the media or device.

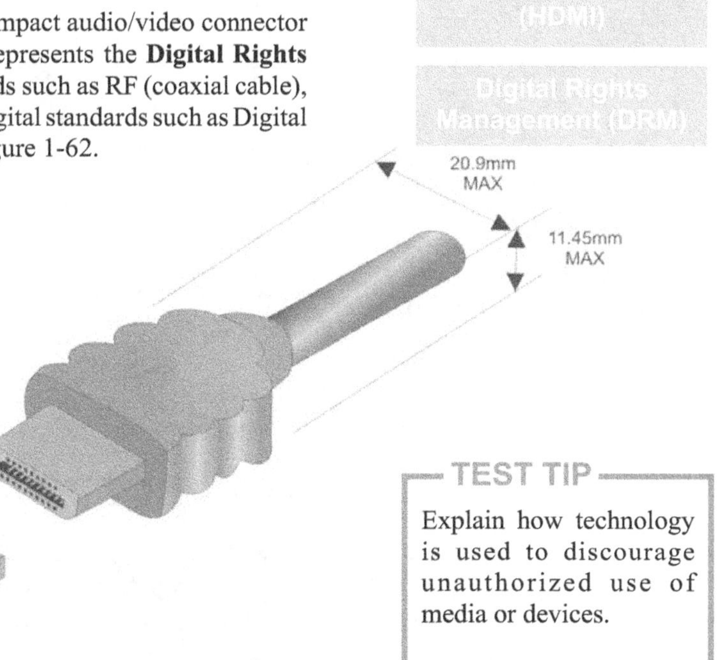

Figure 1-62: HDMI Connector

Know which interface cable can carry digital audio and video between multiple devices.

HDMI connects DRM-enforcing digital audio/video sources such as a set-top box, an HD DVD disc player, a Blu-ray Disc player, a personal computer, a video game console, or an AV receiver to a compatible digital audio device and/or video monitor such as a digital television (DTV). HDMI began to appear in 2006 on consumer HDTV camcorders and high-end digital still cameras.

Digital Visual Interface

Digital Visual Interface (DVI)

The **Digital Visual Interface (DVI)** is a video interface standard designed to maximize the visual quality of digital display devices such as flat panel LCD computer displays and digital projectors. It was developed by an industry consortium, the Digital Display Working Group (DDWG). It is designed for carrying uncompressed digital video data to a display. It is partially compatible with the High-Definition Multimedia Interface (HDMI) standard in digital mode (DVI-D). The connector for a DVI-D interface is shown in Figure 1-63.

Be able to differentiate between various PC video connection options.

Identify compatibilities between different A/V connection standards.

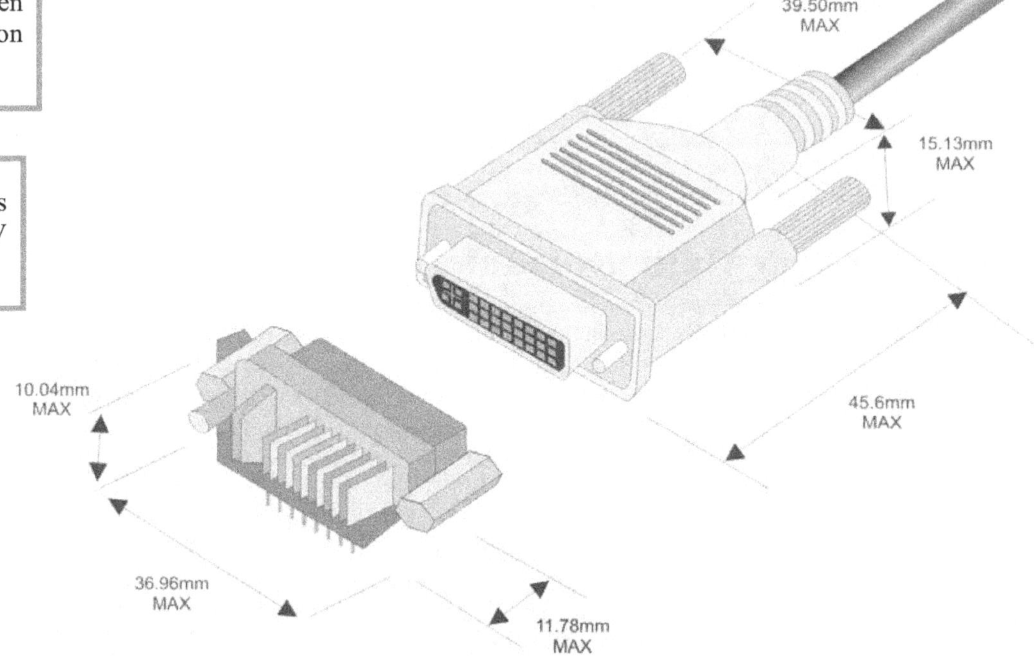

Figure 1-63:
DVI-D Connector

ADVANCED AUDIO/VIDEO DESIGN CONCEPTS

High-Definition Television (HDTV)

Home entertainment systems have become a popular segment of the home technology industry. **High Definition Television (HDTV)** broadcasting and multichannel surround sound speaker systems have created interest in dedicated home theater designs.

While many home entertainment systems utilize a dedicated home audio/video design where all components are installed in a single room, some of the latest technologies in audio/video home entertainment can expand these systems from the home theater room to other areas in the home. This type of design is referred to as a whole-home entertainment system.

A whole-home audio/video system design allows you to see and hear all of the shared video and audio sources in multiple zones throughout the home. Audio equipment in a central location (family room, den, etc.) distributes sound to speakers located in other rooms and areas. Video distribution uses a central video distribution panel, amplifiers, and modulators to distribute high-quality video information to multiple areas in the home.

Audio/Video system design alternatives include both **dedicated** and **distributed** architectures. This terminology is used by contractors to describe the dedicated home theater design option versus a distributed design with audio/video components in multiple areas of the home. Figure 1-64 shows a typical A/V design that places all of the A/V equipment in a room dedicated to home theater functions.

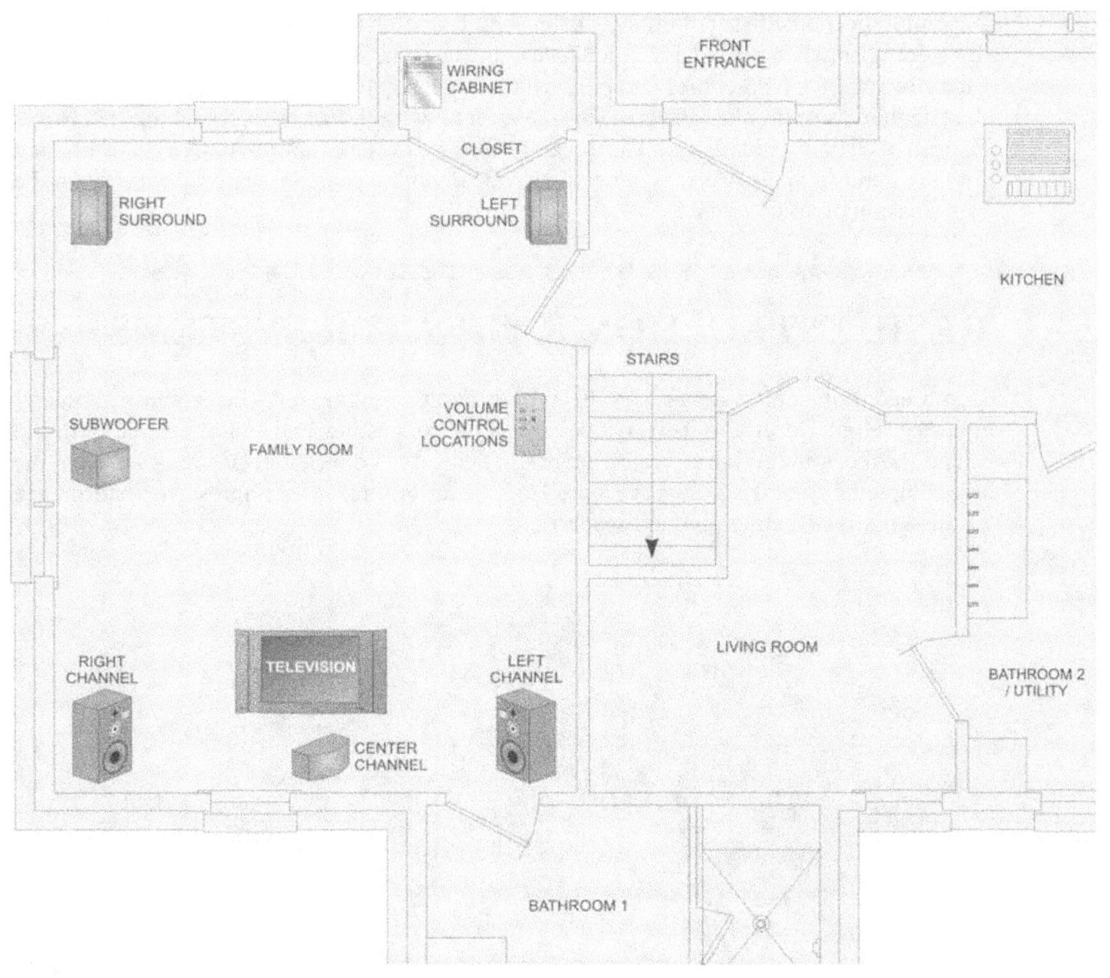

**Figure 1-64:
A Dedicated
Home Theater
Installation**

A dedicated system design:

- Places all audio/video equipment in one location.
- This is the typical design choice for a home theater system where a location such as a den or a custom-designed entertainment center is the single dedicated location for home entertainment.

- This is the preferred design for home theater surround sound speakers and home theater source equipment such as large-screen, high-definition television (HDTV) systems or video projection systems.

While many home entertainment systems utilize a dedicated home audio/video design where all components are installed in a single room, some of the latest technologies in audio/video home entertainment can expand these systems from the home theater room to other areas in the home. This type of design is referred to as a whole-home entertainment system.

Whole-home audio/video system design requires a distribution system to provide multimedia entertainment throughout the home. Components such as digital versatile disk (DVD) players, amplifiers, TVs, video cassette recorders (VCRs), and personal video recorders (PVRs) all require different types of connectors and cables with different formats. Whole-home designs may also include the connection and remote control of home appliances, telecommunications devices, and security systems in addition to home audio and video systems.

A communications/distribution system in which one channel is used for signaling, and different channels are used for voice/data transmission, is known as **Common Channel Signaling (CCS)**. CSS comes into play within a home networking environment, whenever the telephone system must interface or interoperate with it. For example, in the PSTN one channel of a communications link is typically used for the sole purpose of carrying signaling for the establishment of a clear connection. The remaining channels are used entirely for the transmission of voice or data.

RESIDENTIAL HOME THEATER SYSTEMS

Home entertainment systems have become a popular segment of the home technology industry. High Definition Television (HDTV) broadcasting and multichannel surround sound speaker systems have created interest in dedicated home theater designs. An example of a completed system is shown in Figure 1-65, which optimizes the impressive performance of these integrated entertainment systems.

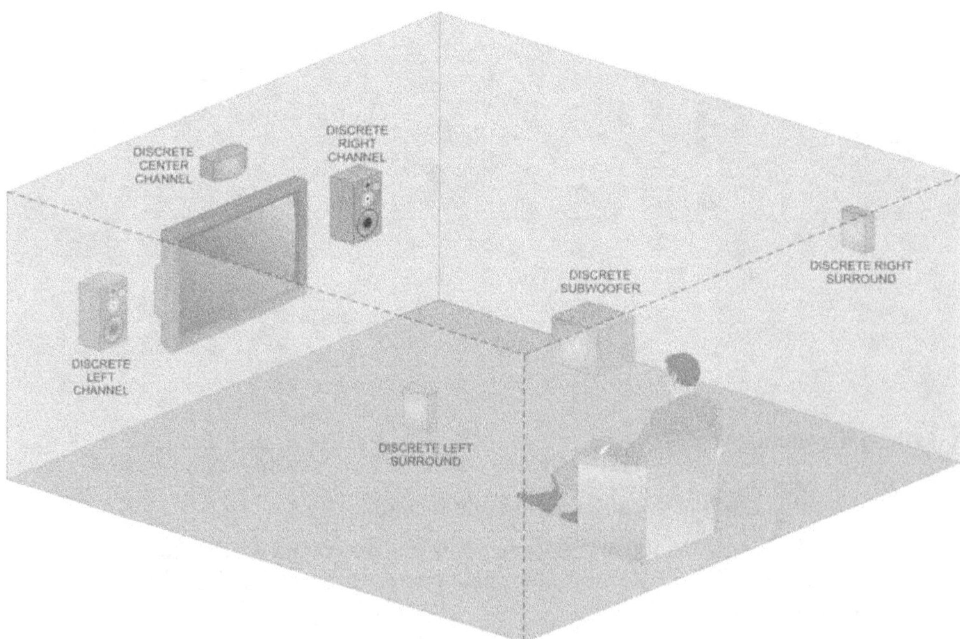

**Figure 1-65: Typical
Home Theater System**

Surround Sound

Surround sound is a term used to describe a type of audio output in which the sound appears to "surround the listener" by 360 degrees. Surround sound systems use three or more channels with speakers located in front and behind the listener to create a surrounding envelope of sound and directional audio sources. The term surround sound has become popular feature of home theater systems.

Surround sound speaker systems use five speakers placed in special locations in the front and side listening areas and a subwoofer to handle the low-frequency effects channel. Subwoofers are speakers that produce only the lower frequencies in the region of 20–80 hertz.

Surround sound speakers can be mounted in a home theater or any location where a surround sound system installation is practical. The subwoofer is a larger speaker and can be located anywhere without sacrificing performance. This is due to the omnidirectional properties of bass sound waves. Additional information on the various surround sound formats is given later.

Discrete and Matrixed Channels

5.1 surround sound is the most common format for home theater systems. It includes a total of six channels—five full-bandwidth channels with 3-20,000 Hz frequency range for front left and right, center, and left and right surrounds, plus one "low frequency effects" (LFE) subwoofer channel for frequencies from 3-120 Hz.

- **Discrete**: Some channels are considered "discrete"—that means that the sound information contained in each of the available channels is distinct and independent from the others.

- **Matrixed**: Other channels are considered "matrixed"—that means that the sound information in those channels is extrapolated from information in other channels.

- **Lossless**: Most surround formats are compressed so that they're small enough to be stored or transmitted—on a DVD, for example, or in a satellite TV broadcast. But now, some higher-capacity HD DVDs and Blu-ray Discs can hold lossless, uncompressed surround soundtracks, for more detailed audio.

Digital Theater Sound

Digital Theater Sound (DTS) it is a multichannel surround sound format used in both commercial and consumer applications. DTS was created by the company for which the technology was named, Digital Theater Systems (which is now called DTS as well). DTS is a competing technology with Dolby Digital.

Dolby Surround

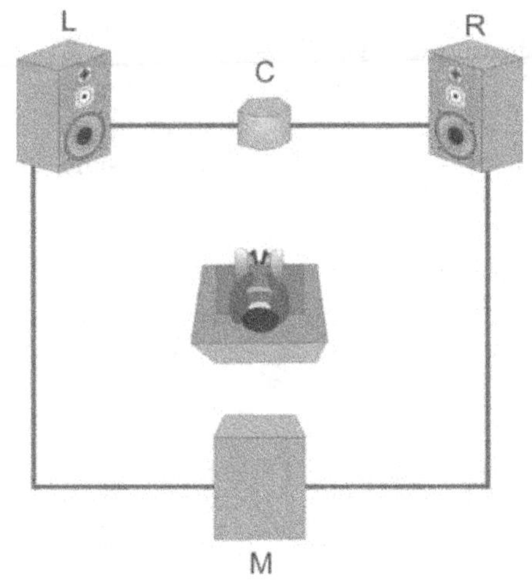

**Figure 1-66:
Dolby Surround
Sound**

Dolby Surround

Dolby Surround is the consumer version of the original Dolby multichannel analog film sound format—Dolby analog and Dolby SR (Spectral Recording). When a Dolby Surround soundtrack is produced, four channels of audio information (left, center, right and mono) surround are matrix-encoded onto two audio tracks. These two tracks are then carried on stereo program sources such as videotapes and TV broadcasts into the home where they can be decoded by Dolby Pro Logic to recreate the original four-channel surround sound experience, depicted in Figure 1-66.

Dolby Digital

Dolby Digital

low-frequency
effects (LFE)

Dolby Digital (formerly known as AC-3) features five discrete Dolby Surround channels (front center, front left, front right, surround left, surround right. The optimum speaker arrangement for implementing this surround sound configuration is shown in Figure 1-67. The 5.1 designation, the standard for DVD sound tracks, is derived from the ability to deliver 5 channels of full-frequency sound, plus a sixth channel for **low-frequency effects (LFE)**. The LFE signal is usually reserved for the subwoofer speaker, or those speakers capable of reproducing low-frequency ranges. The low-frequency effects channel gives Dolby Digital the ".1" designation. The ".1" signifies that the sixth channel is not full frequency, as it contains only deep bass frequencies (3 Hz to 120 Hz.)

TEST TIP

Understand optimal speaker arrangements for various surround sound configurations.

**Figure 1-67:
Dolby Digital
5.1 Surround
Sound**

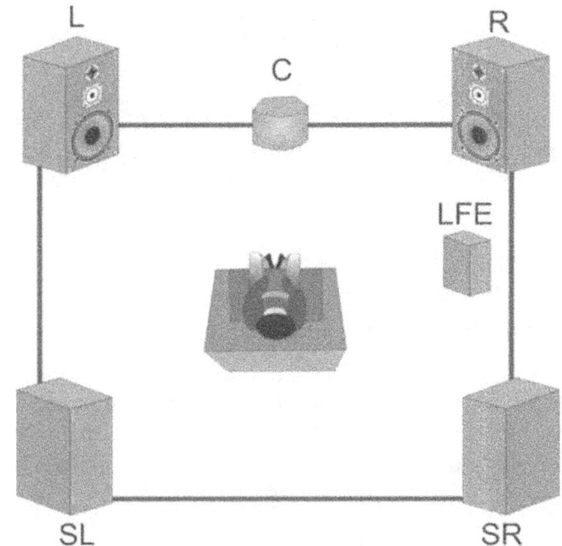

Digital Theater Sound – Extended Sound

Digital Theater Sound – Extended Surround (DTS-ES) is a 6.1 matrixed system that enables cinemas to deliver extreme spatial effects that appear to surround the audience. A DTS-ES decoder creates a back-surround (BS) channel from encoded surround tracks, in typical theater applications feeding a back-surround speaker array (configured as left-back-wall and right-back-wall for stereo operation). The 6.1 configuration is shown in Figure 1-68. The system is compatible with all current extended surround formats and an auxiliary surround channel is also provided for other applications.

During system calibration for a 6.1 surround system, it is important to set the speaker delay correctly for proper time synchronization. Delay settings that are too long may result in an echoed, unnatural surround field, causing unnatural imaging shifts. However, settings that are too short may result in a flat 2-dimensional sound field between the front and rear speakers.

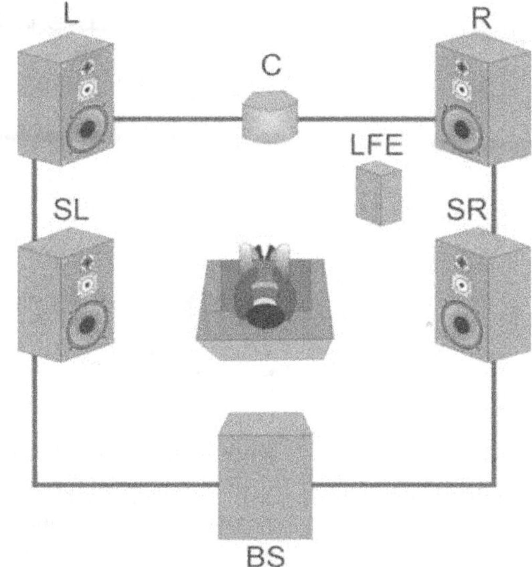

**Figure 1-68:
Surround Sound 6.1**

Dolby Digital Plus, DTS-HD

Surround receivers have had 7.1 outputs for some time, but there was not a discrete 7.1 format with which to use them. Receivers had added a right rear and a left rear instead of the rear center found in a 6.1 system. Again, this did NOT get you discrete 7.1—the two rear back speakers were matrixed from other channels

The increased storage capacity of Blu-Ray and HD-DVD allow both formats to carry 8 discrete channels of audio. The DD Plus and DTS-HD formats are capable of supporting more than 8 discrete channels if future technologies call for them. This audio configuration is depicted in Figure 1-69.

There are two uncompressed versions of these formats—**Dolby TrueHD** and **DTS-HD Master Audio** are lossless. Using higher sampling rates, these two formats ensure that every bit of data is reproduced, ensuring the highest fidelity, or faithfulness to the original possible.

All four formats have been engineered to be somewhat backwards compatible with their respective companies' earlier decoders. While this will allow discs with this encoding to be used on older systems, there are no discs that have these formats encoded on them at the time this training was written.

> **TEST TIP**
>
> Realize the importance of eliminating "speaker delay" in a surround sound system.

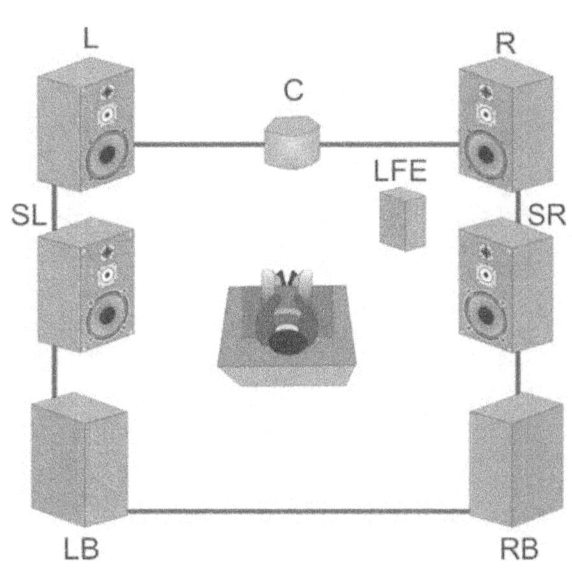

**Figure 1-69:
7.1 Surround
Sound**

Super Audio CD

Super Audio CD (SACD) is an optical audio disc format aimed at providing higher fidelity digital audio reproduction than the standard audio CD. Introduced in 1999, it was developed by Sony and Philips Electronics, the same companies that created the Compact Disc. SACD was in a format war with DVD-Audio, but neither format managed to replace regular audio CDs

MULTIROOM AUDIO/VIDEO DISTRIBUTION SYSTEMS

Multiroom audio/video distribution is defined as the ability to listen to music and video in different rooms/zones at the same time. In addition to multiroom configurations, single source and multisource distribution systems add to the convenience and versatility of audio and video content throughout the home. Figure 1-70 depicts a whole home A/V installation that provides a dedicated home theater as well as a distributed audio system. This arrangement allows listeners throughout the residence to select music from various sources and deliver it to their location.

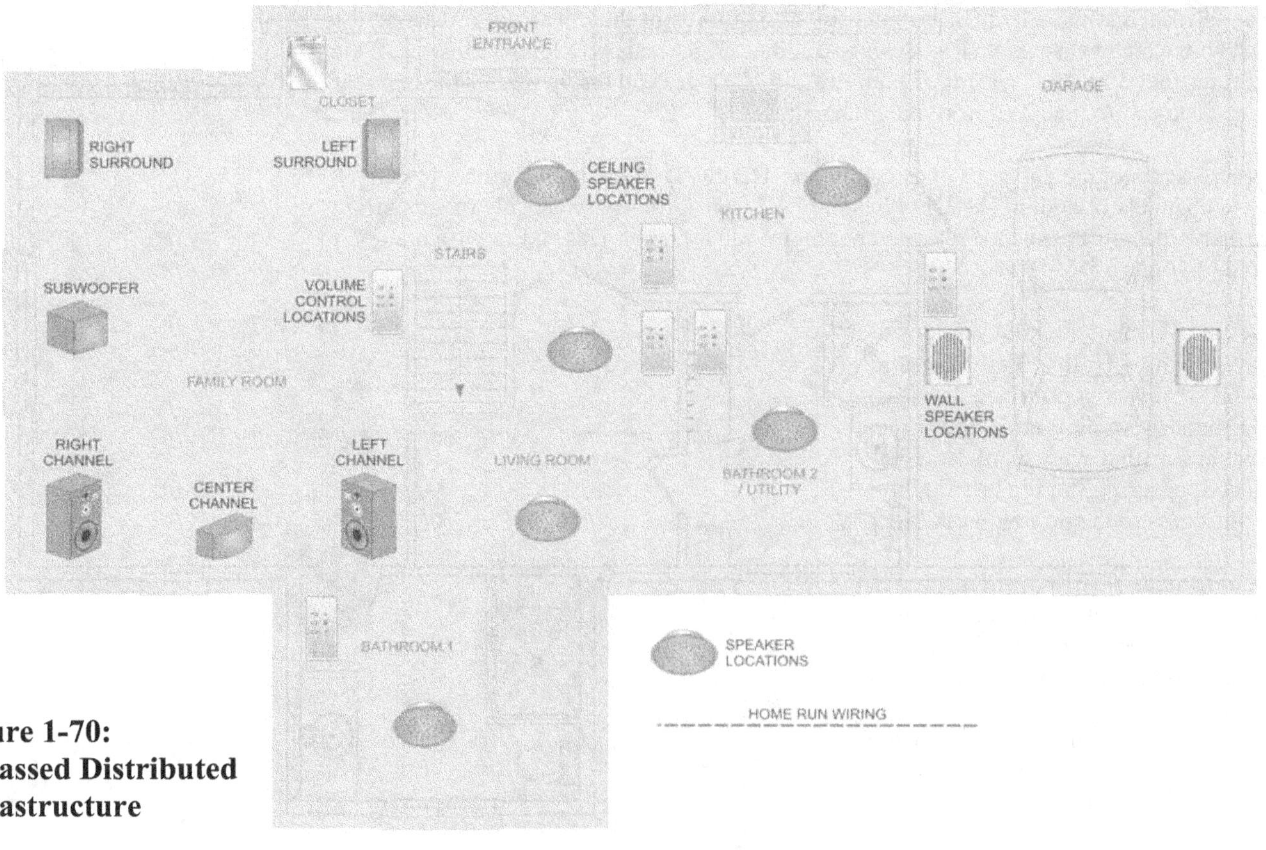

**Figure 1-70:
Bypassed Distributed
Infrastructure**

Multiroom Audio Systems

There are two common types of multi-room audio systems—single-zone systems and multi-zone systems. In a **single-zone**, multi-room system a centralized audio source is used to deliver the same audio information to each area of the home. At any given time, all the speakers in these systems will play the same audio.

The nature of the audio information is determined by whoever has control of the central unit. This unit is typically a **single-source device** such as an AM/FM receiver, a multifunction device such as an AM/FM receiver/DVD player, or an intercom system.

This type of system is typically wired in a home run configuration to multiple speaker/control units in various locations throughout the residence, as depicted in Figure 1-71. At the control unit, the listener can choose to turn the audio off or adjust the volume being produced. However, they do not have the ability to select the audio information being delivered—this is done through the central unit. This type of audio system is the least expensive, and easiest to install multi-room audio system.

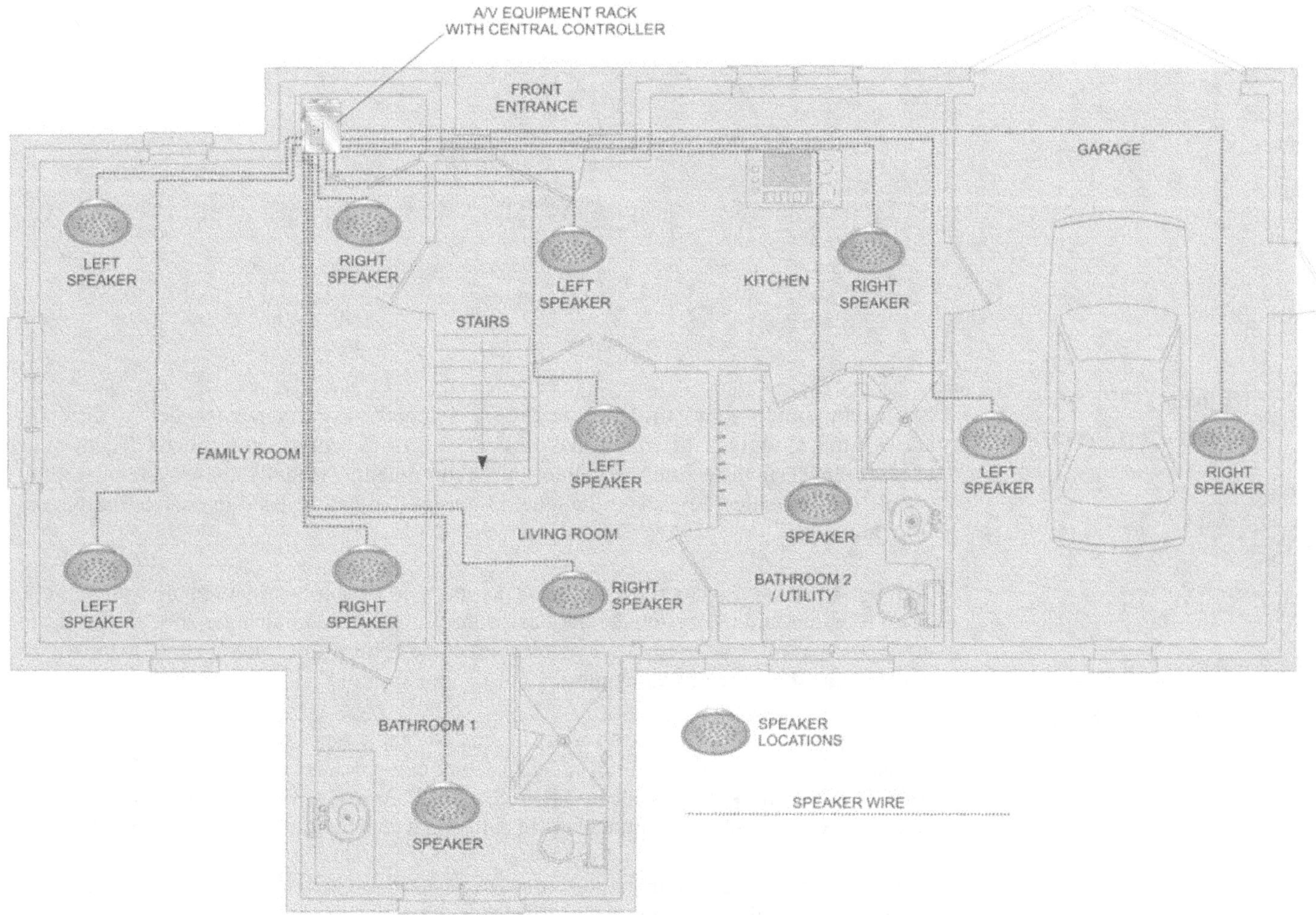

Figure 1-71: A Single Source, Multi-Room Audio System

Multi-zone audio systems provide the ability to listen to music in different rooms/zones at the same time, as shown in Figure 1-72. Audio distribution enables audio source equipment located in a central location to be shared by different users in various locations called **zones**. A zone includes a group of speakers in one area of the home that is controlled by a common controller and plays the same audio signal.

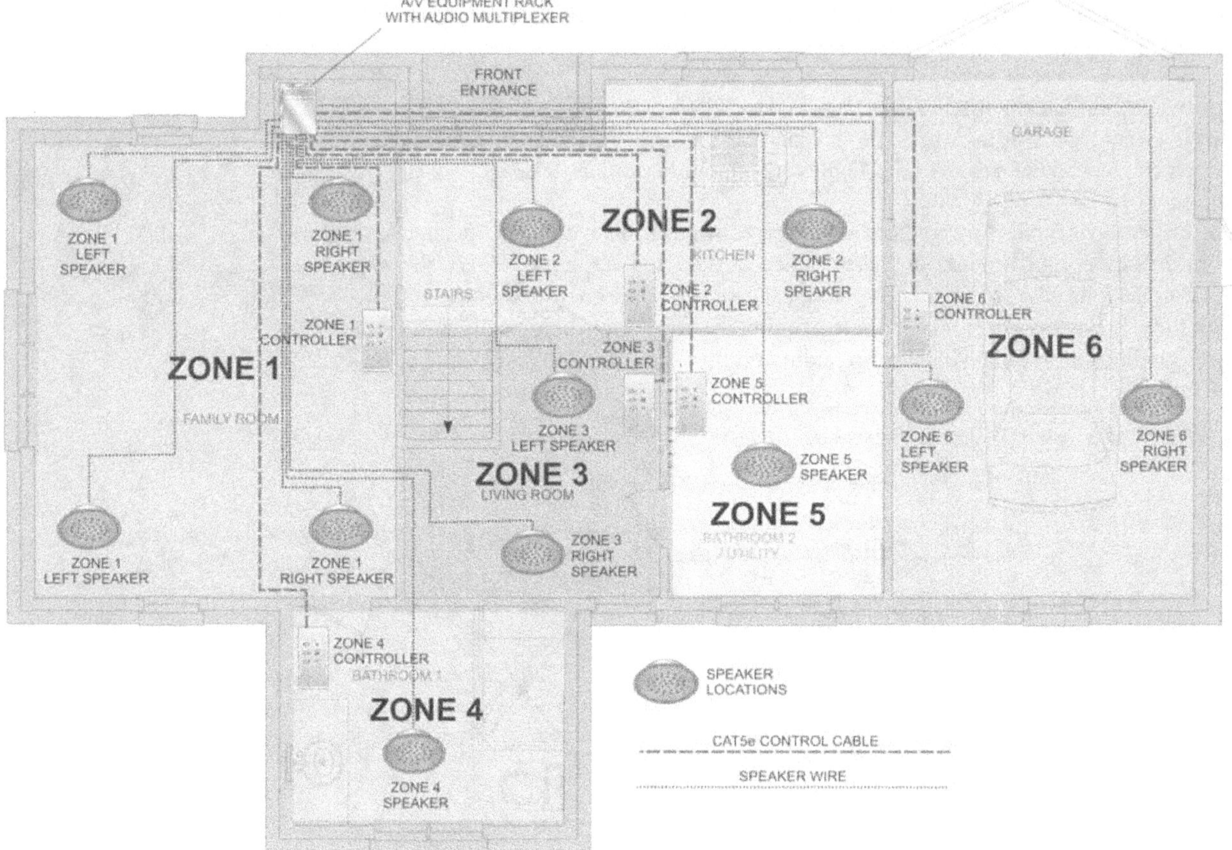

**Figure 1-72:
Multi-Zone Audio**

Multi-room (multi-zone), **multi-source** audio systems can simultaneously deliver different audio output to users in different zones obtained from different sources (tuner, CD player, or cassette player) in the centralized audio source equipment. Typically, these systems employ simple wall-mounted keypads or handheld remotes that can be used to control the system from any zone.

Volume controls and source equipment keypads are usually mounted in wall outlets. They're wired into the system so they can control the sound in a single zone or the entire home. Keypads can be used to select the source and control the volume either manually or with handheld infrared remote units.

Audio distribution systems use a variety of wire and cable types for connecting components including Category 5 UTP cable, speaker cable, and shielded audio patch cables. Audio source equipment is usually located in a "headend" location where all equipment for the audio and video system components is installed in a shared audio/video equipment rack.

> **TEST TIP**
>
> Understand the symptoms associated with open or shorted distribution cabling.

The ability to select different audio sources at the headend location and deliver it to a given zone is provided through an audio **multiplexer**. These devices are the A/V equivalent of a router in the networking environment. Their job is to use programming input from users to channel the correct audio source to the desired audio zone.

As Figure 1-73 illustrates, the multiplexer provides a number of source inputs and a number of zone outputs. A low-end multiplexer may offer as few as four source inputs and four zone outputs. Other, more expensive models may offer more than 8 sources and 16 zones. Some models also have the ability to cascade (operate multiple units together to provide even more source and zone options).

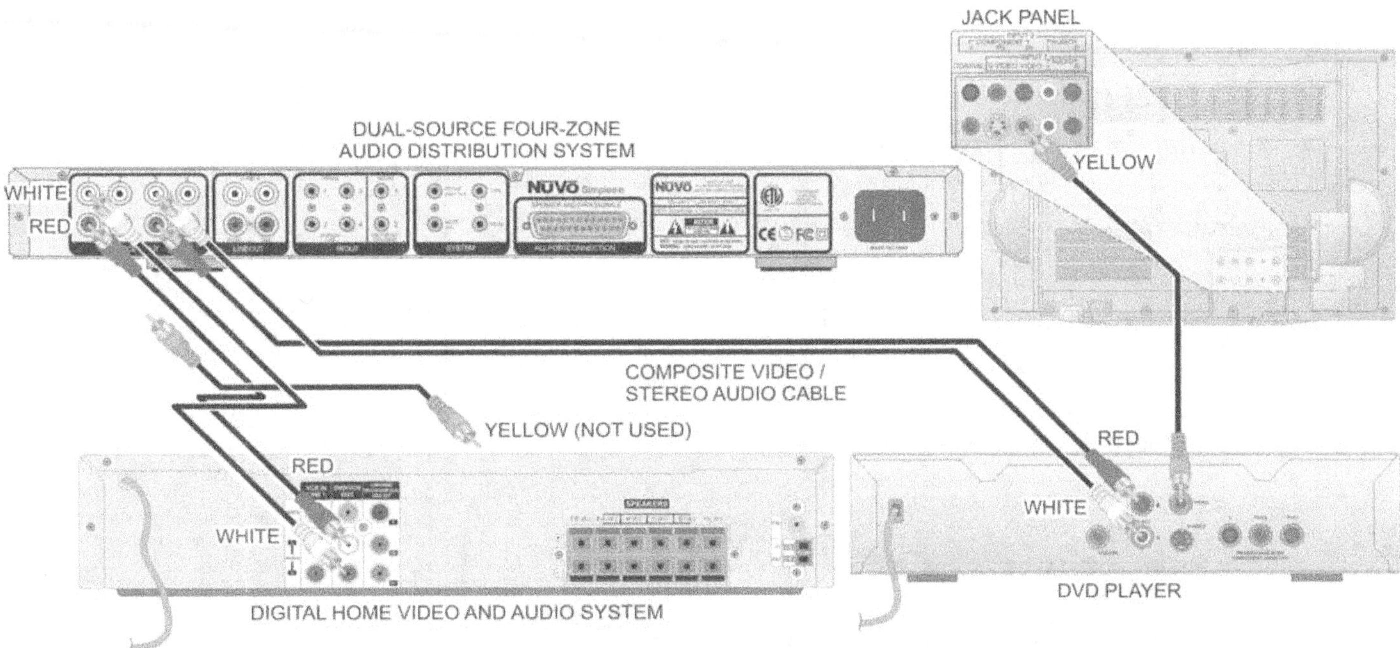

Figure 1-73: An Audio Multiplexer System

For an A/V system designer the key with these systems is to determine what is available and match that to the layout of the home, the listening interests of the customer and the price of implementing the system. For example, if you chose the 4-source, 4-zone model to install in a twelve-room house, several rooms would have to share a zone (i.e., 3 rooms per zone). If some rooms in this example need to be their own zone—such as the master bedroom, the other zones would need to take on more speaker pairs.

In the figure, two audio sources (a DVD player and an AM/FM receiver with built-in DVD player), are connected to the audio inputs of the four-source, four zone multiplexer. The output from this multiplexer model is a multi-pin cable that connects to a master distribution block, depicted in Figure 1-74, designed to fit in a standard residential light switch enclosure.

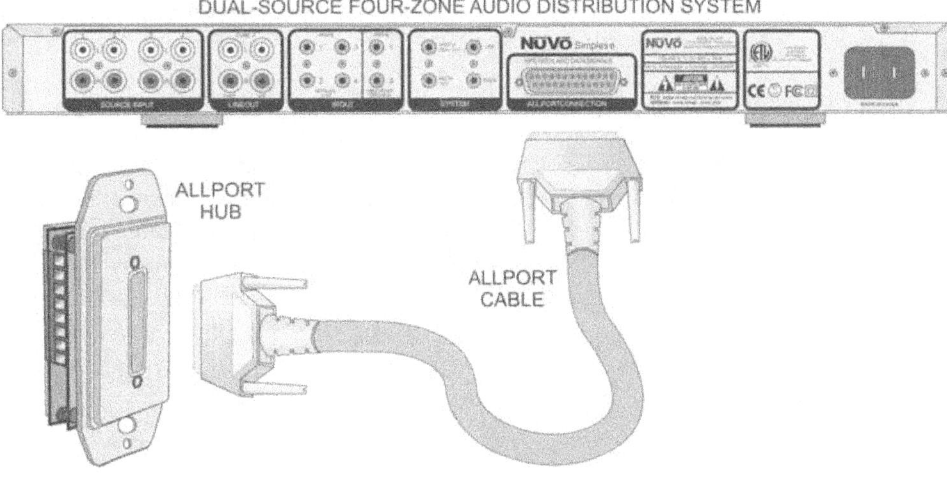

Figure 1-74: Audio Distribution Block

The distribution block provides connectivity in two directions—inbound digital control signals that travel over CAT5/5e cables from the different zone controllers (up to four for this model) and outbound *analog audio signals* that travel over speaker wire to the zoned speakers. This concept is illustrated in Figure 1-75. Some multi-room A/V systems also employ the CAT5/5e cable to return *digital audio signals* to the zones. The digital signal is converted into analog signals to drive the speakers at the speaker locations.

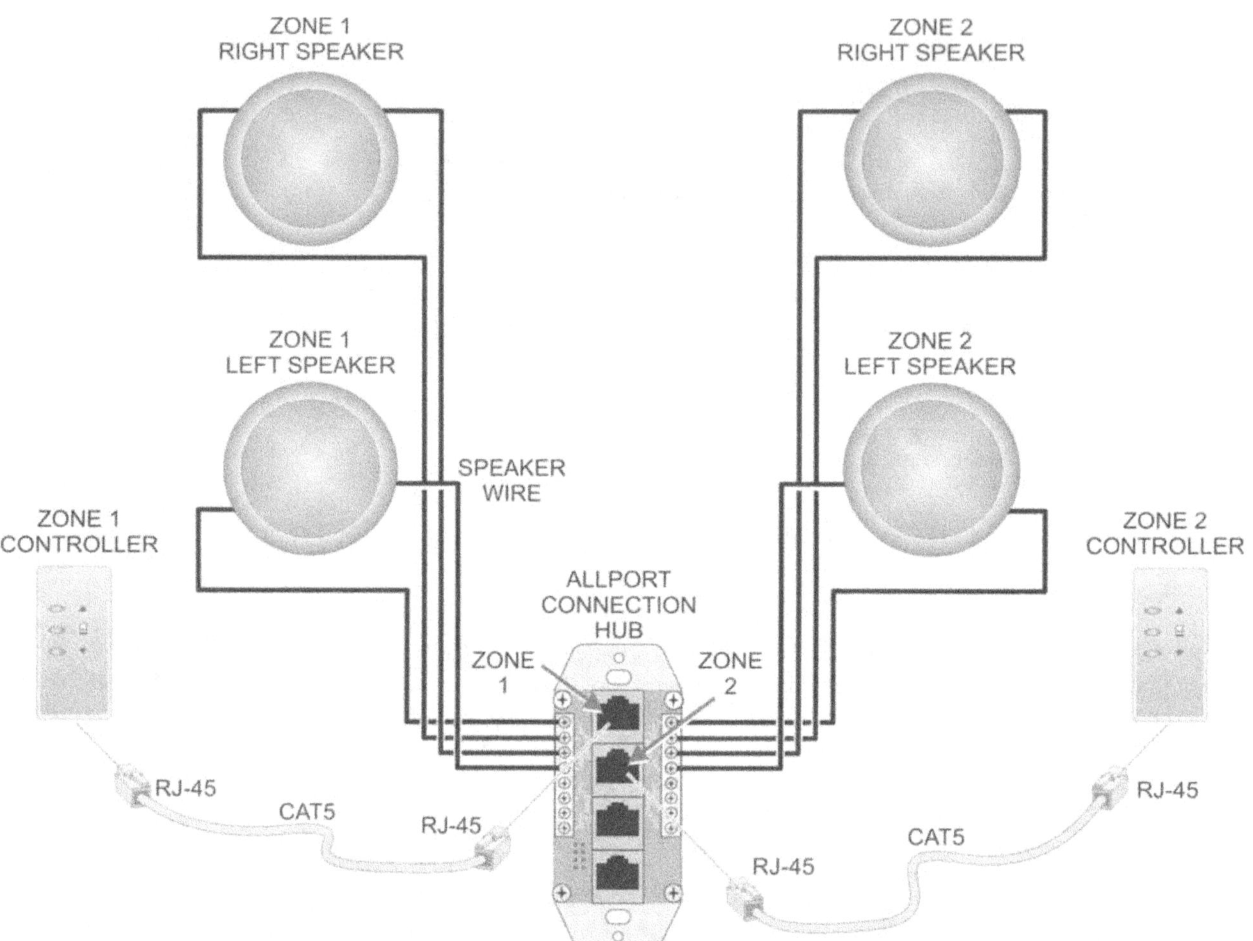

Figure 1-75: Audio Distribution Wiring

The zone controllers (keypads or touchscreens) pass digital information from the zone to the multiplexer. In turn, the multiplexer generates infrared control signals that are applied to the different pieces of source equipment through **infrared (IR) emitters**, as illustrated in Figure 1-76. The emitters are required because nearly all A/V equipment is designed to be controlled by IR remote controls.

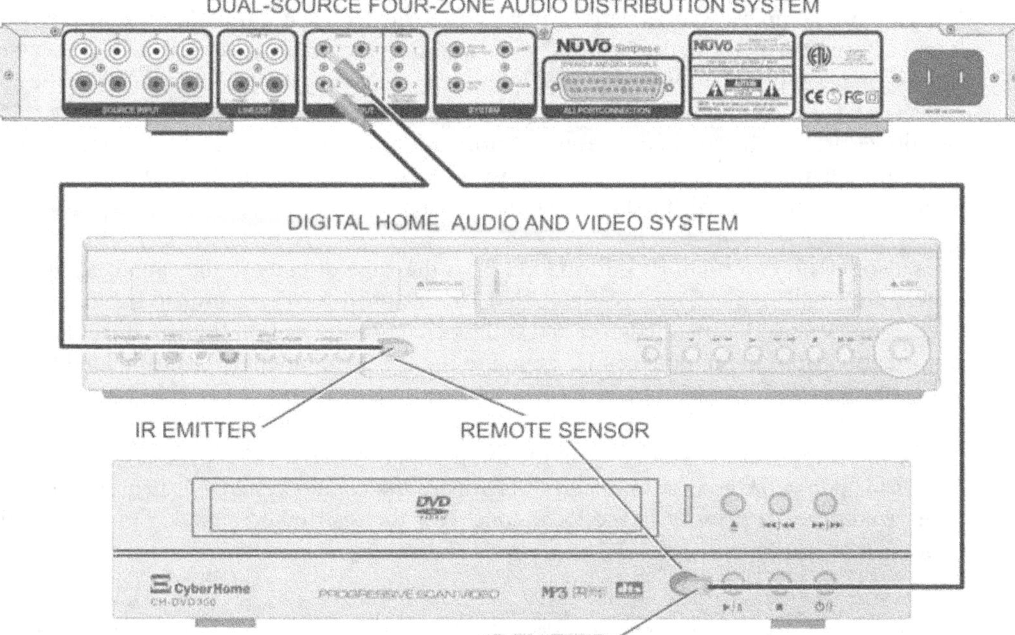

Figure 1-76: Using IR Emitters

The zone controllers typically have an embedded IR receiver that allows inexpensive, programmable IR remotes to be placed in each zone and used to send commands to the system without physically interacting with the controller. Figure 1-77 depicts a typical wall mounted keypad that permits the listener in its zone to select the source that has the audio information they want to listen to and to control the volume level for the zone. This controller also enables the listener to turn all zones off—effectively turning the system off for all zones of the residence, without having to actually visit each zone.

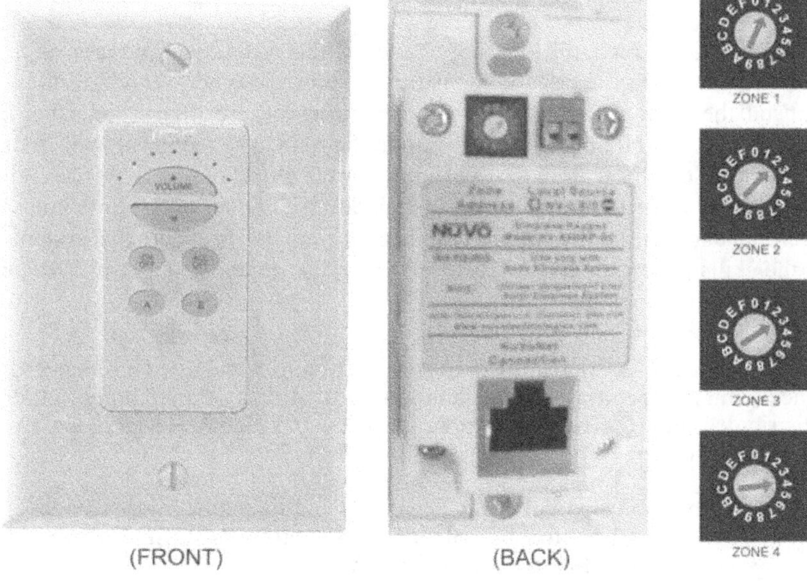

Figure 1-77: A Simple Multi-Room Audio Keypad

Large more complex A/V systems typically use **touchscreen controllers** in place of the simple keypad depicted in the figure. These controllers are necessary to provide the user with the expanded control options associated with such complex systems. For example, if the user selects a *CD stacker* with 200 CDs in it as the source, they will also need to be able to select the title/artist/play list to obtain the audio information they want to hear. This cannot be done through a simple keypad, but the touchscreen controller can interact with the source equipment to provide this type of dynamic information.

It is also possible to route IR information from the zone directly to the source equipment. There are small IR receivers available that are designed for being embedded in the wall. These receivers can be painted over and still receive IR information from an IR remote. The information from the remote can be routed to the A/V equipment over unshielded twisted pair cabling.

Some multi-room audio systems provide for an input from a **local source** (a source device that is located in the zone but not part of the distributed audio system). When this source is energized, it takes priority in the local zone over the distributed signal coming from the central A/V equipment.

MULTIROOM VIDEO DISTRIBUTION SYSTEMS

Distributed video is a system where all kinds of video (and audio) signals from A/V source components are routed to remote display locations. At the remote locations, display devices such as flat panel television receivers and projection systems are used for viewing the information from the A/V sources at the head end.

A distributed video system is desirable when there are multiple A/V sources on site, there are multiple display locations and/or the displays are located a long distance from the source.

Today's home builders recognize the importance of the system integrator as an important partner for work with any residential construction project. When searching for a professional to install the latest A/V technology solutions, installers are often asked if they have undergone the necessary training or certification in the field. The technical skills required by digital home installers differ from those of a traditional wire installer. A certified electronic systems technician must adhere to proper installation techniques and follow industry standards. Such standards and certifications have been formulated to ensure that systems installers have the skills necessary for this type of work. For example, the Multi-Room Audio Standards set by CEA include:

- The ANSI/CEA-2030 Multi Room Audio Cabling Standard. This standard ensures successful multi-room audio installations, providing a detailed blueprint for adding distributed audio to homes. Ample graphics and charts are included along with suggested set-ups.

- The recently published bulletin CEA-CEB17. This publication provides guidance from floor-to-ceiling for the planning, selection, installation, and optimum performance for residential speaker systems.

- The ANSI/CEA-863-A standard. This standard covers the connection color codes for home theater systems.

-

Video Distribution Systems

A model for the design of a home video distribution system is shown in Figure 1-78. Variations on this model are possible depending upon the distance of each cable run, home size, cost, and number of video sources. The figure represents all of the basic features of an average video distribution system.

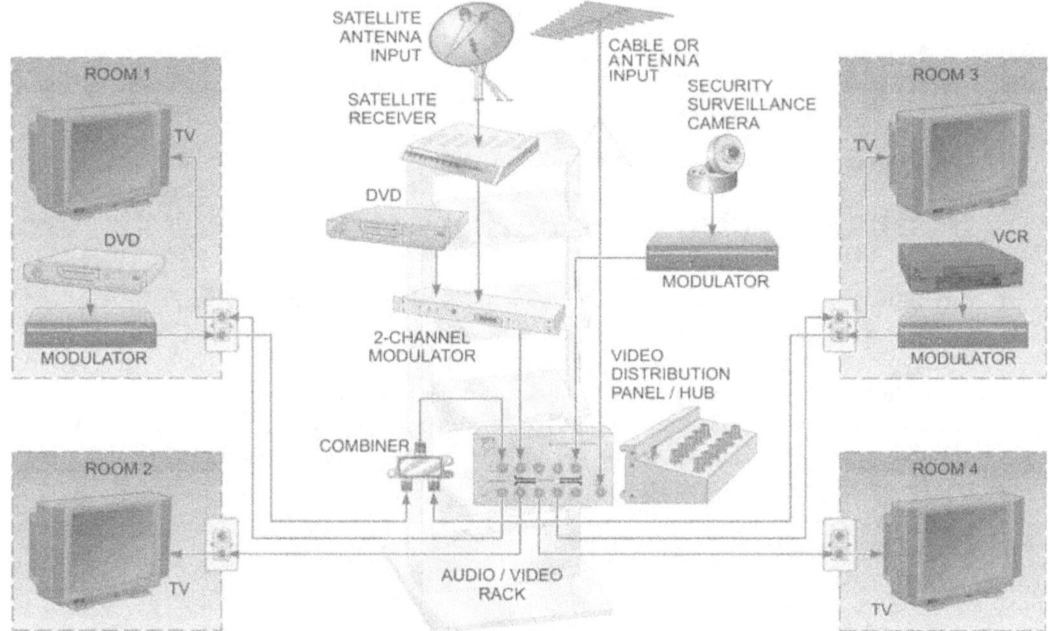

Figure 1-78: Distributed Video System

Features offered by the whole-home video distribution system shown in the figure are as follows:

- The video distribution panel is the central hub for all video signals. It contains amplifiers, splitters, and modulators. It provides the capability to have all home video sources distributed to any television receiver in the home.

- RG6 coaxial cable provides the input and output connections to the video distribution panel. Each room has dual coaxial cable wall outlets. The bottom connector on each outlet takes signals generated in that room from a VCR or video camera and feeds them back to an input on the distribution panel so that they become available to all other TVs in the house.

- The outputs from the video distribution panel provide amplified modulated video inputs to each of the TV receivers in the separate rooms. The amplifiers are adjusted to provide the same video signal quality to each room.

- Modulators convert the video source signal (surveillance camera, satellite receiver, VCR, and DVD player) to an unused television channel for internal broadcasting on the video distribution system. This allows the home users to choose which video sources they wish to view by tuning the television receiver to the desired channel.

A combiner consolidates video inputs from source equipment and connects them to the video input jack on the distribution panel.

There are four methods for distributing video:

- Baseband
- Internet Protocol (IP)
- IEEE 1394
- Radio frequency (RF)

Baseband Video Distribution

Baseband distribution systems are networks that communicate without employing carrier signals. Generally, baseband networks transmit at data rates of less than 10 MHz. These systems typically employ multiple-wire homerun to each display location. These wires are connected to a matrix typically located at the head end. The sources are also wired into the matrix at the head end. The main benefit to this type of system is the higher quality signals. With the multiple signal wires, however, you need a control system to instruct the matrix as to which source signals to route to which outputs.

IP Video Distribution

IP video distribution systems deliver a digital signal on the data infrastructure network. Some of the main benefits are high-quality signals, simple installation with Category 5e or Category 6 unshielded twisted-pair cable and standard RJ45 connections, industry standard infrastructure and the ability to share the same network with other essential business needs, such as a data network and Internet connectivity.

An important aspect of IP digital distribution is that of signal conversions between IP-based and analog-based devices. For example, existing analog CCTV cameras can be interfaced with an analog-to-IP encoder. The cameras can then be added to the IP network through the use of CAT5/5e cabling between the encoder and the computer. This type of conversion permits the use of any IP surveillance software with the system. A centralized server connection will permit access to the system from any PC using a web browser. Likewise, an IP-to-analog video decoder can be used to convert digital camera video back into its analog state for display on a standard security monitor.

IEEE 1394

Another type of digital video interface you will need to remember is the IEEE 1394 interface. It is a very fast external bus standard that supports data transfer rates of up to 400Mbps (in 1394a) and 800Mbps (in 1394b). Products supporting the IEEE 1394 standard go under different names, depending on the company. Apple, which originally developed the technology, uses the trademarked name FireWire. Other companies use other names, such as i.link and Lynx, to describe their IEEE 1394 products. Firewire cables specifically designed to transfer digital video between digital cameras and computers are often referred to as DV cables.

Category 5e/6 Cable

Most video distribution systems used today are designed to be used with 75 Ohm coaxial cables. To transmit these signals over 100 Ohm balanced twisted pair cabling (Category 5e or Category 6) requires a high quality broadband balun (Balanced to unbalanced transformer) at either end of the channel to adapt a 75 Ohm coaxial input using an F-connector to a 100 Ohm balanced output using a modular 8-position connector.

A big advantage of transmitting video over unshielded twisted pair is that it has the capability to transmit many channels simultaneously using only a single pair in a 4-pair cable for broadcast video distribution or two pairs for bi-directional interactive video transmission.

RF Video Distribution

In an **RF video distribution system**, the baseband audio and video outputs of all the source components are modulated onto an RF carrier. These **RF carrier** signals are routed into RF combiners and finally into large RF amplifiers for transmission across the distribution system.

The amplifier or amplifiers used must have adequate gain to supply a clean signal to the most-distant display location. The wiring typically consists of RG-6 or RG-59 shielded coax cable.

In an RF system, coaxial cable home runs to each display location and all the coaxial cables are connected at a head end location with a coaxial splitter(s) and/or combiners, as shown in Figure 1-79. There are also designs based on a single coaxial run with multiple tap points at each display location. In this system, you have a large RF amplifier that can provide enough signal strength to reach the farthest display location, and you have a modulator connected to each source device. The modulator sends the A/V signals to a specific channel that is tuned in at each of the remote display locations.

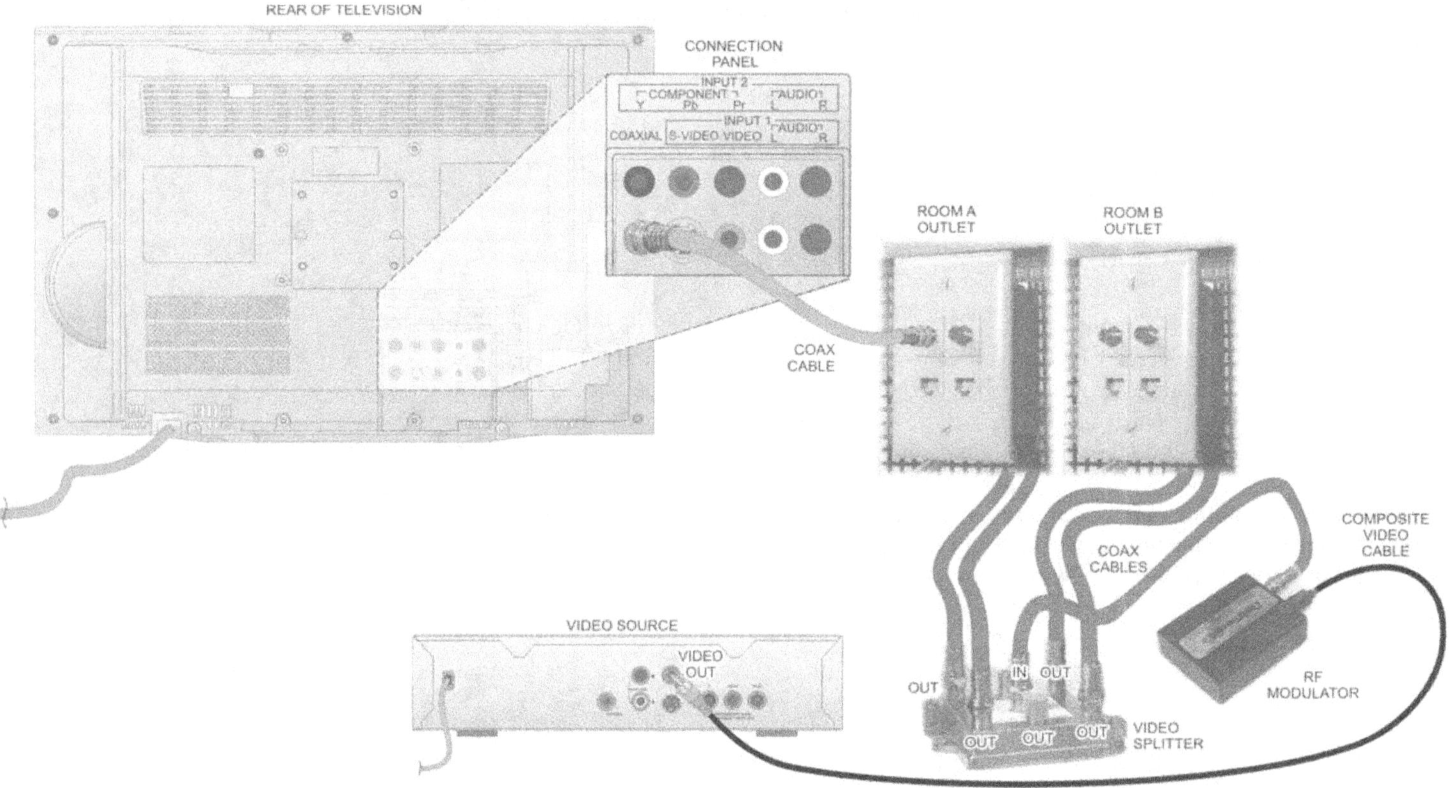

Figure 1-79: RF Video Distribution System

Internal Broadcasting and Video Distribution

Internal broadcasting allows different video sources to be viewed at different locations in the home as shown in Figure 1-80. This is accomplished by tuning a "cable ready" television receiver to an unused channel. "Unused channel" refers to a UHF TV channel for each installed modulator in the range of channels 14 to 78 or CATV channels 65 to 135 that is not used in the local over-the-air or cable broadcast domain. Video sources can then be privately "broadcast" internally. Channels can be selected independently on each television receiver by each user to allow cable channels, a VCR, or a DVD to be viewed on each TV by selecting the correct configured modulator channel.

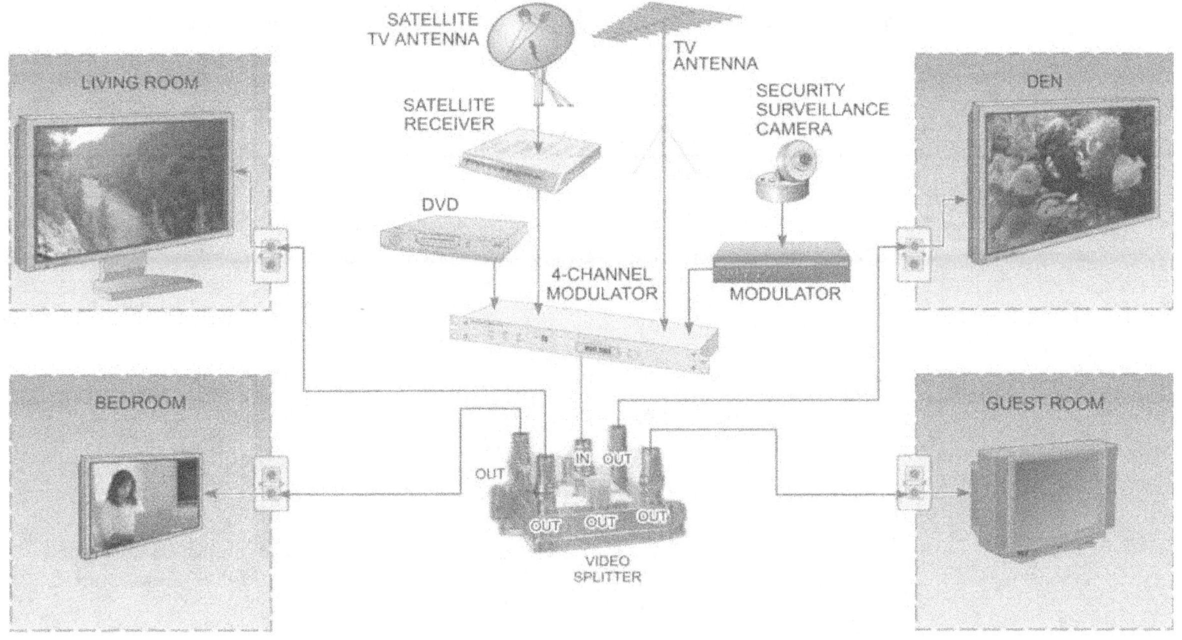

Figure 1-80: Internal Broadcasting

KEY POINTS REVIEW

Review the following key points before proceeding to the Review Questions and Exam Questions sections to make sure you are comfortable with each point. After completing the review, answer the Review Questions that follow to verify your knowledge of the material covered in this book.

- Internal broadcasting allows different video sources to be viewed at different locations in the home.

- The crossover network circuit inside the speaker enclosure divides the audio signal into various frequency bands before sending them to the different speaker drivers.

- Impedance ratings for speakers are not the same as resistance although both are measured in ohms. Resistance is the direct current (DC) resistance and impedance is a value that changes with frequency.

- Surround sound speaker systems use five speakers placed in special locations in the front and side listening areas and a subwoofer to handle the low-frequency effects channel. Subwoofers are speakers that are designed to produce only the lower frequencies in the region of 20–80 hertz.

- Although CRTs are popular and relatively inexpensive, they are limited by economical and technology constraints to display areas of approximately 36 inches.

- CRT monitors and TV displays are specified in size by the diagonal measurement of the tube's front area.

- Personal Video Recorders (PVRs) or Digital Video Recorders (DVRs) are similar to Video Cassette Recorders (VCRs) except that they use a hard disk instead of a magnetic tape as the recording media.

- Digital Light Processing (DLP) is a new display technology. The DLP trademark is owned by Texas Instruments, representing a technology used in projectors and video projectors.

- Satellite receiver/decoders provide the digital decoding and unscrambling of the downlink from a direct broadcast satellite (DBS).

- Front projection utilizes a video projector and a separate pull-down screen to show the projected image, similar to a movie screen.

- Component video is the format that provides the best color reproduction when connecting home theater products.

- Composite video is a video format where color (chrominance), luminance (brightness), and synchronization signals are mixed into a single radio carrier wave.

- A DBS satellite antenna location must have a clear view of the southern sky. The antenna distance above the ground is not important.

- The Digital Visual Interface (DVI) is a video interface standard designed to maximize the visual quality of digital display devices such as flat panel LCD computer displays and digital projectors.

- The High-Definition Multimedia Interface (HDMI) is a compact audio/video connector interface for transmitting uncompressed digital streams.

- Blu-ray Disc (also known as Blu-ray or BD) is a high-density optical disc format for the storage of digital information, including high-definition video (HDTV).

- Video formats used for consumer electronic products are classified according to the number of horizontal lines of resolution contained in each frame. A higher number of lines produce a progressively higher quality picture.

- Dolby Digital (formerly known as AC-3) is a 5.1-channel surround sound format, and is the standard for DVD sound tracks.

- A big advantage of transmitting video over unshielded twisted pair is that it has the capability to transmit many channels simultaneously using only a single pair in a 4-pair cable for broadcast video distribution or two pairs for bi-directional interactive video transmission.

REVIEW QUESTIONS

The following questions test your knowledge of the material presented in this book.

1. What type of system offers centralized control of the A/V system as well as storage and management of the video and audio information?

2. What type of circuit inside the speaker enclosure divides the audio signal into various frequency bands before sending them to the different speaker drivers?

3. Explain how Ohm's law is used to show the relationship between voltage, current and resistance in an electrical circuit.

4. When must all TV broadcast stations be upgraded to be able to broadcast HDTV programming?

5. What are two common types of multi-room audio systems?

6. What type of distribution allows different video sources to be viewed at different locations in the home?

7. What are the names for the three input jacks used for component video information?

8. Describe the difference between discrete and matrixed sound channels.

9. What is the difference between direct view and reflected view displays?

10. What types of speakers overcome some of the deficiencies found in dynamic cone speakers?

11. What are the three existing analog world television broadcast standards in use today?

12. What is the functional purpose of a home theater personal computer?

13. What are the three types of MP3 players?

14. What are the two HDTV broadcast standards that have been adopted by the commercial television broadcast industry?

15. What are the current display technologies behind rear-projection television, having for the most part replaced CRT projectors?

EXAM QUESTIONS

1. What type of design for a home theater system places all audio/video equipment in a single location?
    a. A distributed system
    b. An allocated system
    c. A single-channel system
    d. A dedicated system

2. What high-density optical disc format is used for the storage of HDTV digital information?
    a. component video
    b. Blu-ray
    c. RGB
    d. composite video

3. What is the unit of measurement for speaker impedance?
    a. Watts
    b. Amps
    c. Volts
    d. Ohms

4. What is the purpose of an LFE channel in a Dolby home theater surround
    sound system?
    a. It contains deep bass frequencies (3 Hz to 120 Hz).
    b. The LFE channel adds emphasis to the left and right channels.
    c. It provides additional imaging for loud sounds.
    d. It suppresses the intermodulation distortion in the stereo amplifier.

5. What device provides the ability to select different audio sources at the
    headend location and deliver it to a given zone?
    a. diplexer
    b. multiplier
    c. multiplexer
    d. combiner

6. What type of video display technology is limited in size to approximately
    36 inches by economical and technical constraints?
    a. CRT
    b. Plasma
    c. LCD
    d. Projection

7. What is the aspect ratio for HDTV displays?
    a. 4:5
    b. 8:10
    c. 3:4
    d. 16:9

8. What type of video display is referred to as "micro-projection" or
    "micro-display" technology typically applied in projection TV?
    a. LCOS
    b. DLP
    c. CRT
    d. Plasma

9. What is the storage capacity of a Blu-ray disc?
    a. 40 MB
    b. 300 MB
    c. 50 GB
    d. 130 GB

10. All DTV formats are transmitted using _______ compression.
    a. JPEG
    b. MPEG-2
    c. ATV
    d. MPEG-4

LAB PROCEDURES

Home Audio/Video System Design

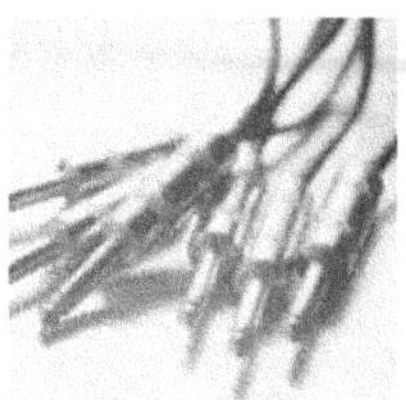

**Audio/Video
Fundamentals**

OBJECTIVES

1. Inventory and validate the audio/video lab equipment.
2. Describe the functions of components included in the audio and video design.
3. Interconnect home theater components.
4. Verify the operation of the home theater system.

RESOURCES

1. Marcraft Audio/Video Experiment Panel and Frame
2. DVD player system
3. Digital home audio and video entertainment system
4. Sound effects CD

TOOLS

1. Flat-tip screwdriver, precision head
2. Flat-tip screwdriver, standard 6"
3. Philips screwdriver, standard 6"

DISCUSSION

Speaker Locations

Speaker systems are designed for installation in a variety of home locations including outdoors. Small bookshelf speakers, flush-mount in-wall speakers, and weather-resistant speakers may be utilized around the home. Some speakers are designed to be partially buried in the ground or disguised as rocks for pool or patio use. In modern home theater installations, surround sound is the preferred system configuration for speaker locations.

Surround Sound Speaker Location

Locating the speakers is an important step in planning a home theater system. A typical home theater layout with *surround sound speakers* is shown in Figure 1-1.

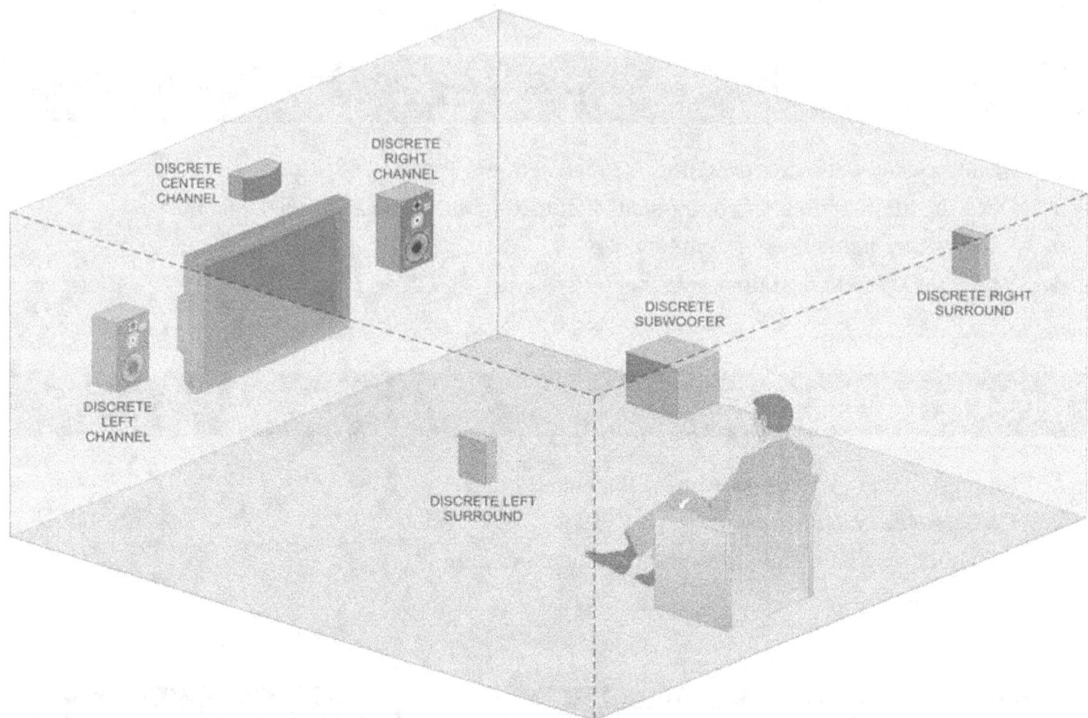

Figure 1-1:
Surround Sound
Speakers

The Dolby Digital surround sound "5.1" format provides up to five discrete (independent) channels (center, left, right, surround left, and surround right, giving it the "5" designation) of full-frequency effects (from 20 Hz to 20,000 Hz). The plan places two speakers in the front of the room to the left and right of the listener. The third speaker should be located in the center and in front of the listening position. A pair of surround sound speakers, one on each side and slightly above and behind the audience, completes the surround sound effect. A sixth channel, dedicated for low-frequency effects (LFE), is reserved for the subwoofer speaker.

Television Viewing Location

The placement of the television is dependent upon several issues. It should be located where the light from exterior sources can be controlled. The size of the room will determine the maximum TV screen size as well as the optimum viewing distance. The minimum recommended viewing distance for a 40-inch television screen is 10 feet.

Video Displays

Video displays accept and process video information from various sources and provide the output data as visual images. Examples of video sources found in home audio/video systems include desktop computers, laptop computers, TVs, closed-circuit TV cameras, camcorders, video projection systems, and high-definition television (HDTV) receivers.

Video displays are manufactured using one of the three basic display technologies listed below:

- Cathode Ray Tube (CRT) display technology
- Flat panel display technology
- Video projection technology

Source Equipment

Source equipment includes all components in a home audio/video system that play back, decode, capture, or process information and output the analog signals for processing by speakers or video display systems.

Audio/video distribution systems are used to select various components for playing music or routing different video channels from source equipment, such as satellite receivers and DVD and VCR playback equipment, to different rooms. Source equipment for a home audio/video system includes the following components:

- DVD players/decoders
- Laser disk player
- Cable TV set-top box/decoder
- Off-the-air TV antennas
- Compact disk (CD) players
- Audio/video receivers
- Tuners

- Amplifiers
- Personal Video Recorders (PVRs)
- Video cassette recorders
- Audio/video distribution systems
- Satellite receivers/decoders
- Modulators

Audio/Video Rack Locations

The best location for the audio/video rack in a home theater room is in the front of the viewing area. This simplifies the layout for wiring between the source equipment, the TV, and the speakers. For a whole-home design, the audio/video equipment rack is typically located near the main audio/video distribution panel.

DVD Players

These units usually play DVDs recorded for a particular region. The region mark on the back of the player must correspond to that on any DVD discs you wish to play. Units sold in North America are coded for Region 1. Most DVD units support CD-R/RW discs recorded with MP3 or JPG files in a particular format. DVD players may have one or more of the following output connectors: Stereo audio out and coaxial digital outputs. Basic units may not include input connectors.

Digital Home Audio and Video Entertainment System

This type of system is usually more flexible than a basic DVD player. The system usually comes with all of the necessary components to provide basic home entertainment. Some units can receive AM and FM radio frequencies as well as play various DVD/CD formats. Output connectors are still standard for audio and video. There may also be input connectors that allow other equipment to be used in conjunction with the devices included in the system. If the system is equipped for surround sound, there should be a speaker connection area, that identifies each speaker's room placement.

Video Signal Formats

Video connections on the rear panel of video consumer electronic devices are configured with different jacks to connect television receivers, DVDs, VCRs, PC monitors, laserdisc players, and camcorders. This section describes the various types of formats for both audio and video connections.

Video formats include:

- Component video
- Composite video
- S-Video
- RGB video

Audio formats include:

- Analog audio
- Digital audio

PROCEDURES

Connecting the Video Equipment

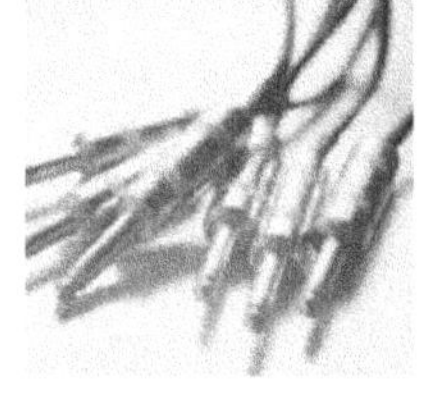

Audio/Video Fundamentals

NOTE: The TV monitor cables and power adapter should already have been connected during the panel mounting process. Therefore, these cables only need to be connected to the media player equipment.

1. Refer to Figure 1-2 during the connection of the video equipment.

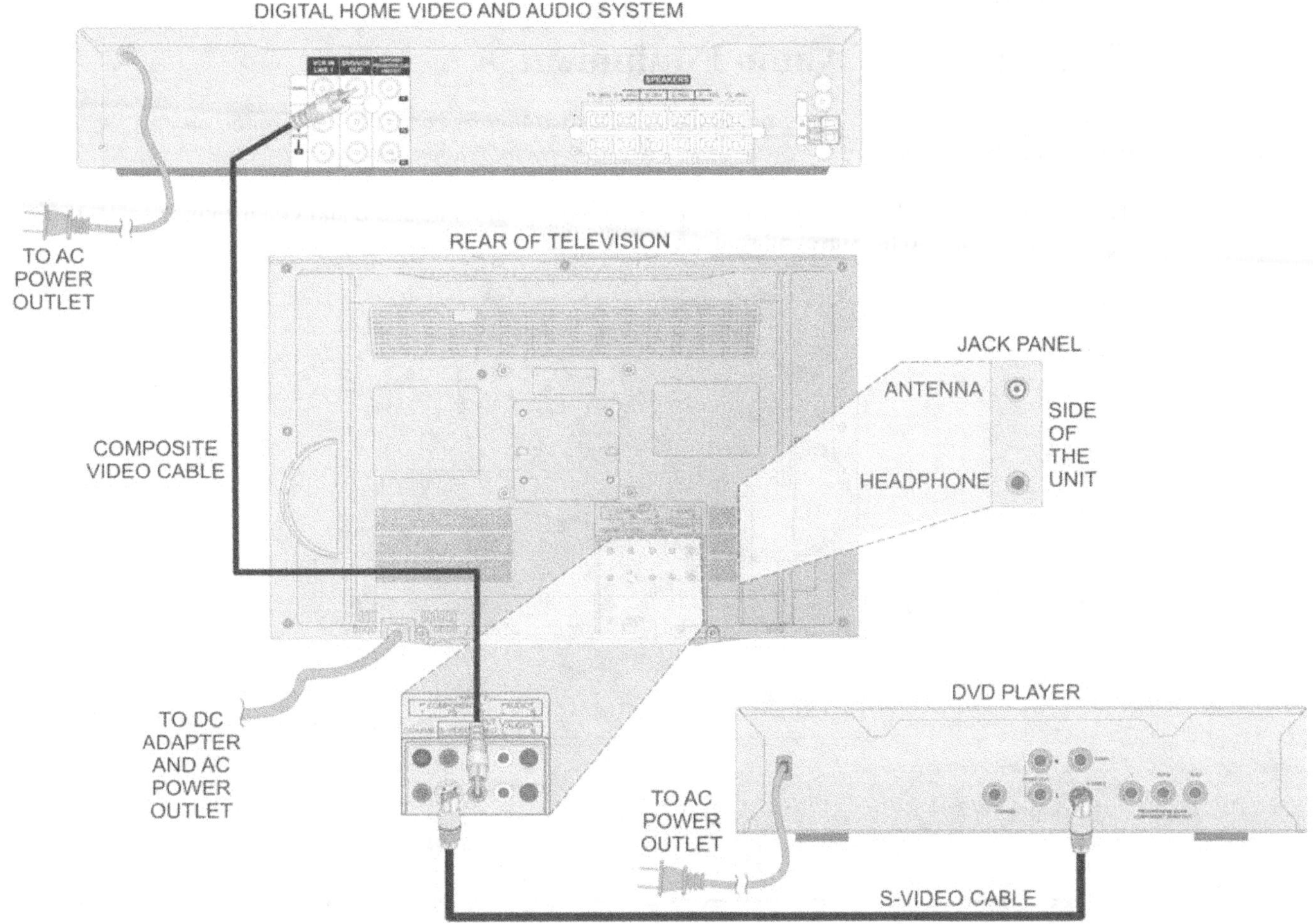

NOTE: The manufacturer has supplied the DVD unit with a three-connector cable that is a composite video/stereo audio cable. One end of the cable (specifically the video connector — Yellow) should already be connected to the flat screen TV. This was performed during the original mounting process. Use the remaining yellow connector to connect the video source from the DVD player to the flat screen TV.

Figure 1-2: Video Connection Diagram

2. Connect the S-Video's cable from INPUT1 of the TV monitor to the DVD player's S-Video's output jack.

3. Connect the Composite Video cable from INPUT 2 of the TV monitor to the digital home audio and video entertainment system's Composite output jack.

4. Plug the TV monitor, DVD player and digital home audio and video entertainment system into a 110-volt ac source.

NOTE: Depending on the manufacturer of the Digital Home Video/Audio system and DVD Player or Blu-ray Disc player the connections using (S-Video/Component/Composite), cabling may need to change. Consult the manufacturer's manual of the Electronic devices you are installing for the proper connections and setup procedures.

Connecting the Audio Equipment

NOTE: The TV monitor has built-in speakers, which will not be connected. The audio outputs for the video equipment will be connected to a sound distribution system in later procedures.

1. Refer to Figure 1-3 during the connection of the digital home audio and video entertainment system's audio outputs to the surround sound speaker system.

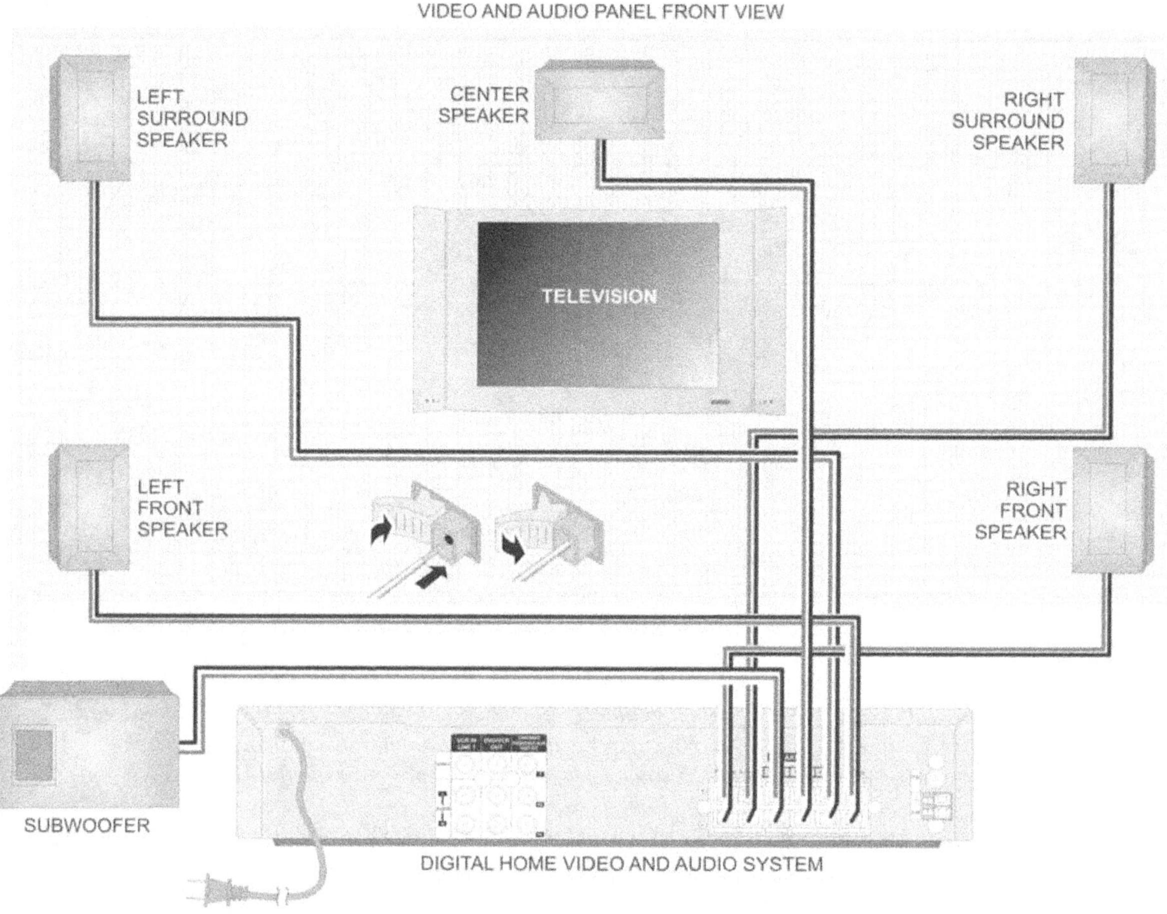

Figure 1-3: Audio Surround Sound Connection Diagram

NOTE: If the five surround sound speakers are premounted on the panel, then the color-coded wire should already be connected to the speakers. The speakers might also be color-coded. If they are not color-coded, then the connection points should be marked as "+" and "-" or colored as red and black, meaning positive and negative respectively. The installer at this point should determine which wire runs connect to which connection points on the equipment. If a system is installed that is not color-coded, the wires should be
labeled at both ends with the intended usage.

2. Using the manufacturer's instruction manual, and following any color-coding and markings on the equipment, connect the five surround sound speakers.

3. Complete Table 1-1.

Table 1-1: Audio Surround Sound Color-Code Connection Scheme

	Example SUB Wire Color	SR Wire Color	SL Wire Color	R Wire Color	C Wire Color	L Wire Color
Top Connector	*Violet*					
Bottom Connector	*Black*					
Abbreviation Definition	*Subwoofer*					

4. Supply power to the digital home audio and video entertainment system.

5. Supply power to the LCD TV monitor.

6. Put the sound effects CD in the digital home audio and video entertainment system.

NOTE: The unit should start playing the effects sequentially. If not, check the digital home audio and video entertainment system's DVD/VCR switch to be sure the selection is set to DVD.

7. At this point you can allow the system to play the tracks sequentially, or push the stop button. Pushing the stop button will cause the unit to display the programming menu.

8. If you prefer to program selected tracks over sequential play, refer to the "Playing Music CDs" section of the instruction manual supplied with the digital home audio and video entertainment system.

NOTE: When choosing specific tracks to play, choose tracks that will demonstrate good surround sound examples. Stationary object sounds might not demonstrate the complete function of the surround sound system. Examples of good surround sounds are trains, plains, objects leaving, and objects passing at high speed.

9. When you have verified the speakers as functional, return the sound effects CD to its case.

10. Turn off the LCD TV monitor and the digital home audio and video entertainment system.

Feedback

LAB QUESTIONS

1. How many independent channels are provided by the Dolby Digital surround sound "5.1" format?

2. What are the names of the independent channels?

3. What is the speaker placement for the Dolby Digital "5.1" format plan?

4. What is the full frequency spread for the Dolby Digital surround sound "5.1" format?

5. What issues are addressed concerning placement of the television?

6. What equipment is included in the definition of source equipment?

7. What are the various types of formats for video connections?

8. What are the various types of formats for audio connections?

Installing a Multizone Audio System

OBJECTIVES

1. Interconnect the components of a multizone audio system.
2. Identify connection points between the audio and video components and the multizone audio system.
3. Verify the operation of the multizone audio system.

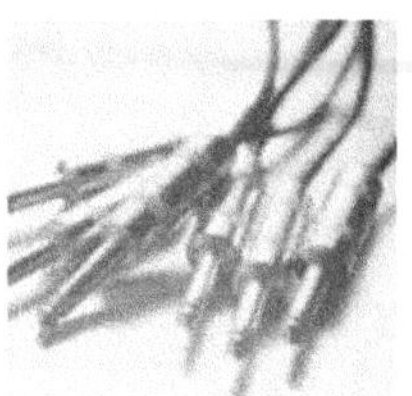

Audio/Video Fundamentals

RESOURCES

1. Marcraft Audio/Video Experiment Panel and Frame
2. Dual-source, four-zone audio distribution system
3. DVD player system
4. Digital home audio and video entertainment system
5. Two RJ-45 cable connectors
6. 3 ft of UTP CAT5 cable

TOOLS

1. Wire strippers
2. Wire cutters, 6"
3. RJ-45 cable connector crimp tool
4. Cable tester scanner

DISCUSSION

Audio Distribution System

An audio distribution system shares some of the features of a video distribution system. Audio distribution requires that audio source equipment located in a central location be shared by different users in various locations called zones. A zone consists of a group of speakers in one area of the home. Audio sources can be designed with either a single-zone or a multizone distribution plan. A single-zone design enables all zones to hear the same audio source at any time. Multizone design permits different users in different areas of the home to listen to different sources independently.

Volume controls and source equipment keypads are usually mounted in wall outlets. They're wired into the system so they can control the sound in a single room or in the entire home. Keypads can be used to select the source and control the volume either manually or with handheld infrared remote units.

Audio distribution systems use various wire and cable types for connecting components including Category 5 UTP cable, speaker cable, and shielded audio patch cables. Source audio equipment is usually located in a "headend" location where all equipment for the audio and video system components is installed in a shared audio/video equipment rack.

Selection of audio source equipment and controls depends upon the size of the home, the number of rooms, and the number of zones; however, a basic design includes the following types of audio source equipment:

- Speakers mounted in each zone using wall, ceiling, outdoor, or bookshelf mounts
- A preamplifier for multi-source input and output switching and source component selection
- An AM/FM radio tuner and amplifier, either an integrated amplifier/tuner design or separate components, also providing input and output switching for other sources such as CD players and tape
- A CD player with either a single tray or multiple CD selection
- Wall-mounted volume controls and component selection keypads

Tape cassette players/recorders have all but disappeared as an audio source; however, they are often included in home designs to provide an easy and economical means for recording audio sources from FM radio broadcasts. Digital Audio Tape (DAT) offers superior quality compared to analog tape players/recorders.

Remote Access

Many remote access systems are available for use with home audio/video. Most components have their own wireless infrared remote controllers. Universal handheld infrared controllers can be programmed to work with a number of different components.

Zoning Distribution

As discussed earlier, zoning distribution is an option for the home entertainment system where different sources can be selected for different areas of the home. A multiroom, multisource system can deliver audio and video throughout the home with the ability to play different sources (tuner, CD player, or cassette player) in different areas simultaneously. Any source can be controlled in any area of the home by a simple wall-mounted keypad or handheld remote. In a single-zone system, all speaker pairs will play the same audio at any given time. This downsized system is least expensive and simplest to install.

PROCEDURES

Connecting the Speakers for the Four Zones

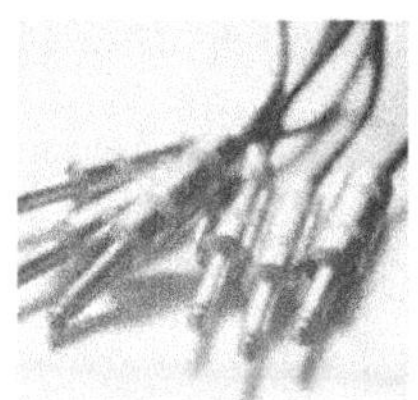

NOTE: If the speaker wire for the Zone 1 speaker is already installed and stripped, proceed to step 2. Also the AllPort Connection Hub may already be installed from previous classes so check with your Instructor first.

Audio/Video Fundamentals

1. Strip about ¼ **inch** of the outer jacket from the speaker end of the Zone 1 speaker wire.

2. Locate and remove the AllPort Connection Hub from the box labeled Multiport Connection on the panel. Refer to Figure 2-1 and the manufacturer's instruction manual while connecting the four zone speakers.

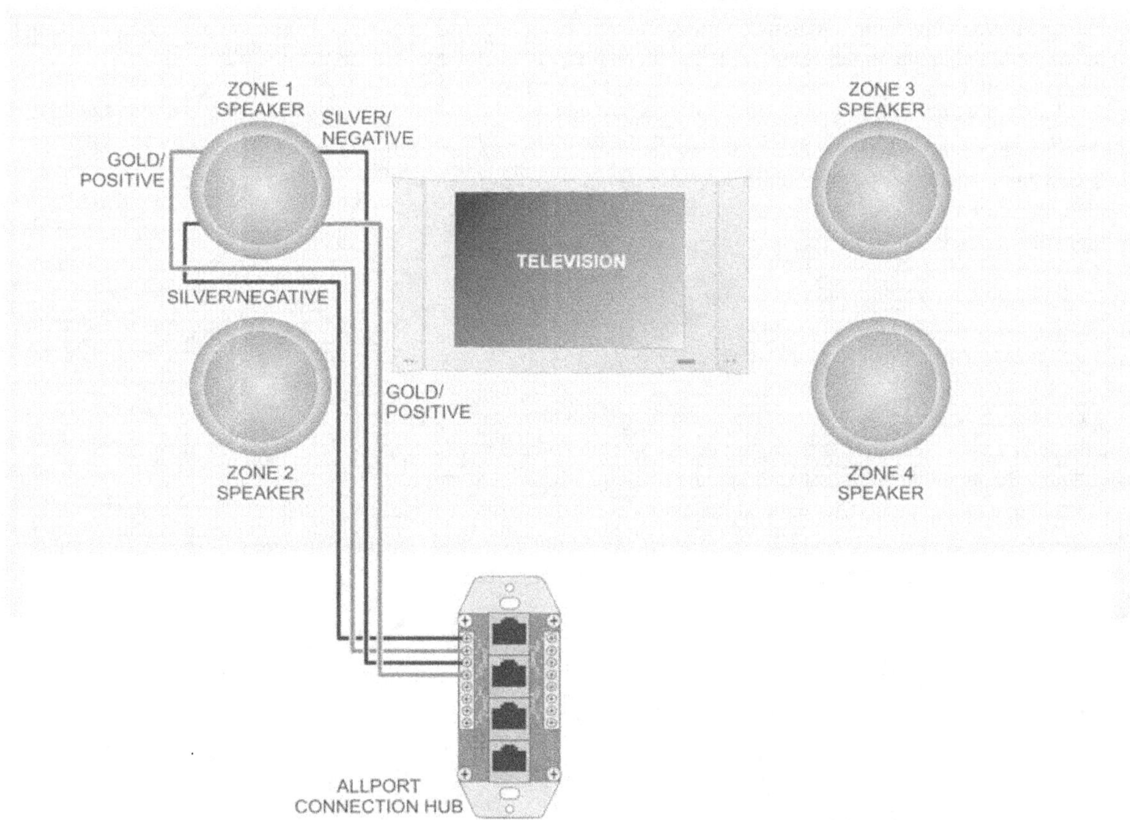

Figure 2-1: Speaker Connection Diagram for the Dual-Source, Four-Zone Audio Distribution System

3. Following the polarity markings on the devices ("+" for the red contact gold wire and "–" for the black contact silver wire), connect the left side of the Zone 1 speaker to the Zone 1 left channel connection block of the AllPort Connection Hub.

4. Repeat step 3 for the Zone 1 speaker's right side.

5. Tighten the four screws on the Zone 1 connection block of the AllPort Connection Hub.

NOTE: The speaker wires for the remaining Zones (2, 3, and 4) are already connected to the speakers. Follow the color-coding or polarity + and - for the speakers and the AllPort Connection Hub, while performing the remaining connection steps.

6. Connect the left and right speaker wires from the Zone 2 speaker to the AllPort Connection Hub Zone 2 connection block. Tighten the four screws.

7. Connect the left and right speaker wires from the Zone 3 speaker to the AllPort Connection Hub Zone 3 connection block. Tighten the four screws.

8. Connect the left and right speaker wires from the Zone 4 speaker to the AllPort Connection Hub Zone 4 connection block. Tighten the four screws.

Connecting the Two Audio Sources

1. Using the manufacturer's instruction manual and referring to Figure 2-2, make the audio connections between the two audio devices and the dual-source, four-zone audio distribution system.

NOTE: The RCA video connector (yellow) may already have been connected to the flat screen TV during the mounting process. Disconnect the S-video from flat screen TV's Input 1. Only the yellow RCA plug should be connected to Input 1 of the flat screen TV.

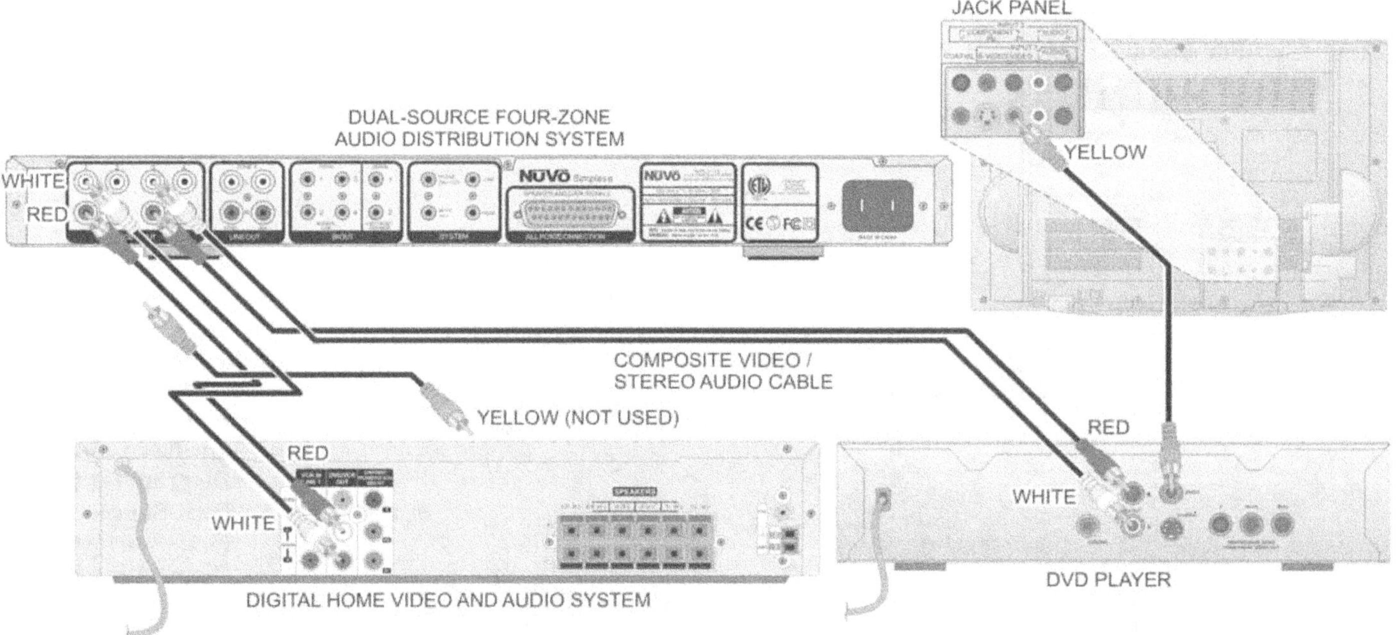

Figure 2-2: Connection from the Two Audio Sources to the Dual-Source, Four-Zone Audio Distribution System

2. Install the digital home audio and video system as the Source 1 input for the dual-source, four-zone audio distribution system.

3. Install the DVD player as the Source 2 input for the dual-source, four-zone audio distribution system.

WARNING

Connecting a *video* signal cable to any of the ports on the dual-source, four-zone *audio* distribution system can cause irreversible damage to the unit.

Connecting AM/FM Radio Antennas

1. Connect to AM loop antenna (Supplied) to the AM antenna connections on the back of your home theater system. See Figure 2-3.

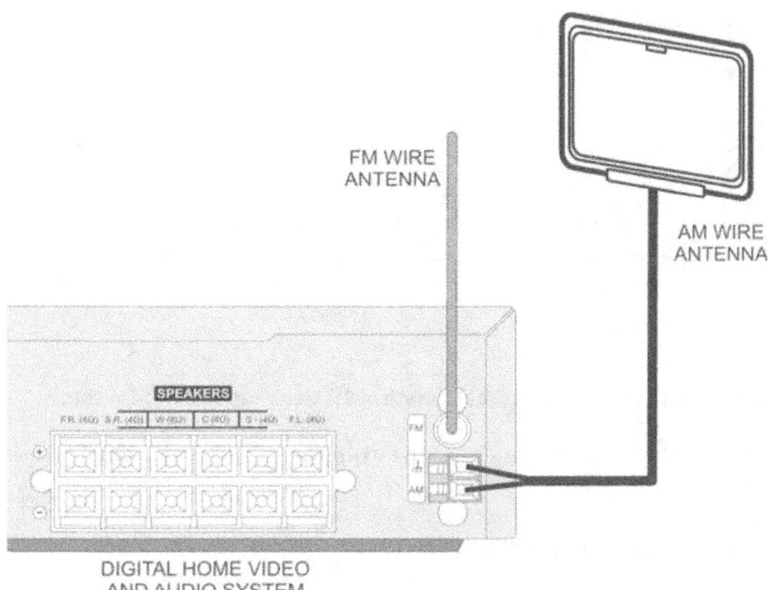

Figure 2-3: Connecting the AM/FM Radio Antennas

2. Connect the FM wire antenna (supplied) to the FM antenna jack on the back of your home theater system.

NOTE: To prevent noise interference, keep the AM loop antenna away from a DVD+Hi-Fi or other de-vices. Make sure that you fully extend the FM wire antenna. After connecting the FM wire antenna, keep it as close to horizontal as possible.

Installing the IR Emitter(s) with Feedback LED

1. Plug one of the IR emitters into the number 1 IR Out port (Normal Use 38khz) of the dual-source, four-zone audio distribution system.

2. Run this IR emitter to the front of the digital home audio and video system.

3. Do not remove the protective paper from the adhesive on the IR emitter.

4. Refer to Figure 2-4 for the location of the remote control IR detector on the digital home audio and video system.

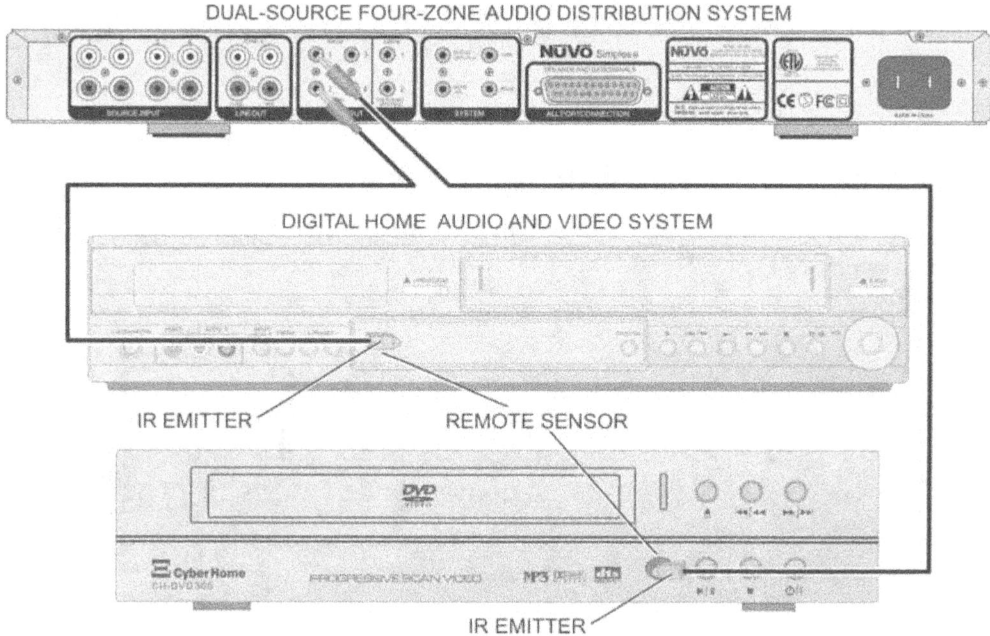

**Figure 2-4:
Installing the IR
Emitters**

5. Gently place the IR emitter over the remote control IR detector and hold it there.

6. Plug the second IR emitter into the number **2 IR Out** port (Normal Use 38khz) of the dual-source, four-zone audio distribution system.

7. Run this IR emitter to the front of the DVD player.

8. Do not remove the protective paper from the adhesive on the IR emitter.

9. Refer to Figure 2-4 for the location of the remote control IR detector on the DVD player.

10. Gently place the IR emitter over the remote control IR detector and hold it there.

Installing the RJ-45 Connectors

NOTE: The cables may already be installed from previous classes. Check with your Instructor first as they may have you build the cables separate from installation starting with step 5.

1. Insert one end of a 4-foot piece of CAT5 cable through a rear access hole in the single-gang plastic box reserved for the Zone 1 keypad.

2. Insert the remaining end of this CAT5 cable into one of the rear access holes in the single-gang plastic box under the flat-screen TV.

3. Insert one end of a 4-foot piece of CAT5 cable through a rear access hole in each of the single-gang plastic boxes reserved for Zones 2−4 keypads.

4. Insert the free ends of these CAT5 cables into each of the remaining rear access holes in the single-gang plastic box under the flat-screen TV.

5. Review Figure 2-5 to determine the pin-out configuration of the 568A CAT5 wiring scheme being used to make the RJ-45 interface cable.

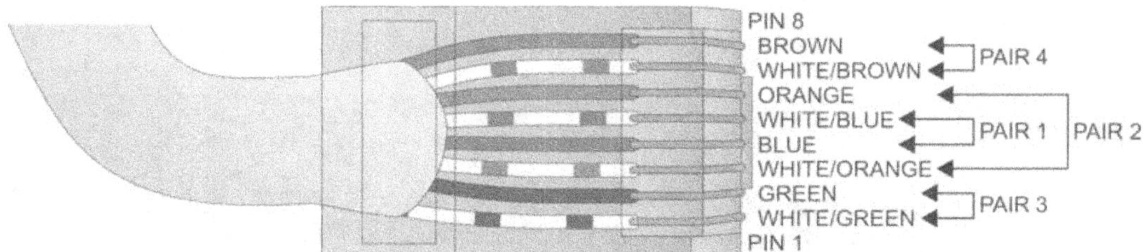

Figure 2-5: 568A CAT5 Wiring Scheme

6. Remove 1/2 inch of the outer jacket from both ends of all five 3-foot pieces of CAT5 cable.

7. On one piece of cable (Zone 1) arrange the conductors so that the color configuration is the same as that shown in Figure 2-5.

8. Insert the conductors into the RJ-45 connector.

9. Check to ensure the color arrangement is still correct.

10. Insert the connector and cable into the crimper.

11. Terminate the connector.

12. Repeat steps 7−11 for the remaining cable end for Zone 1.

13. Plug both ends of the terminated RJ-45 cable into the cable tester scanner.

14. Slide the OFF/ON switch to the **ON** position.

15. Push the **AUTO/MANUAL** button so that the cable tester is in the **AUTO** mode.

16. Complete Table 2-1 for cable 1.

Table 2-1: RJ-45 Cable Connection Test Results

1	2	3	4	5	6	7	8	G
Cable 1: 1st RJ-45 Connector Pin								
Cable 1: 2nd RJ-45 Connector Pin Match								
Cable 2: 1st RJ-45 Connector Pin								
Cable 2: 2nd RJ-45 Connector Pin Match								
Cable 3: 1st RJ-45 Connector Pin								
Cable 3: 2nd RJ-45 Connector Pin Match								
Cable 4: 1st RJ-45 Connector Pin								
Cable 4: 2nd RJ-45 Connector Pin Match								
Cable 5: 1st RJ-45 Connector Pin								
Cable 5: 2nd RJ-45 Connector Pin Match								

17. Turn off the cable tester scanner.

18. Have the instructor verify the results of Table 2-1 before continuing.

19. Repeat steps 7−17 for the four remaining 3-foot pieces of CAT5 cable, using the applicable entries in Table 2-1.

Configuring the Keypad

1. Review the section of the manufacturer's instruction manual that pertains to the Keypad and Zone setup settings for the wall-mounted keypads shown in Figure 2-6.

**Figure 2-6:
Rear View of the
Keypad and Rotary
Switch Settings**

2. Complete Table 2-2 by putting a check mark in the appropriate column to indicate whether the switch is set to its Zone.

Table 2-2: Zone 1 Keypad Rotary Switch Settings

Zone 1 Keypad Rotary Switch Settings			
Switch	1	2	3	4
Set				

3. Set the rotary switch on the **Zone 1** keypad to match the configuration recorded in Table 2-2.

4. Connect the **CAT5 RJ-45** connector from the **Zone 1** single-gang plastic box to the keypad. See Figure 2-7.

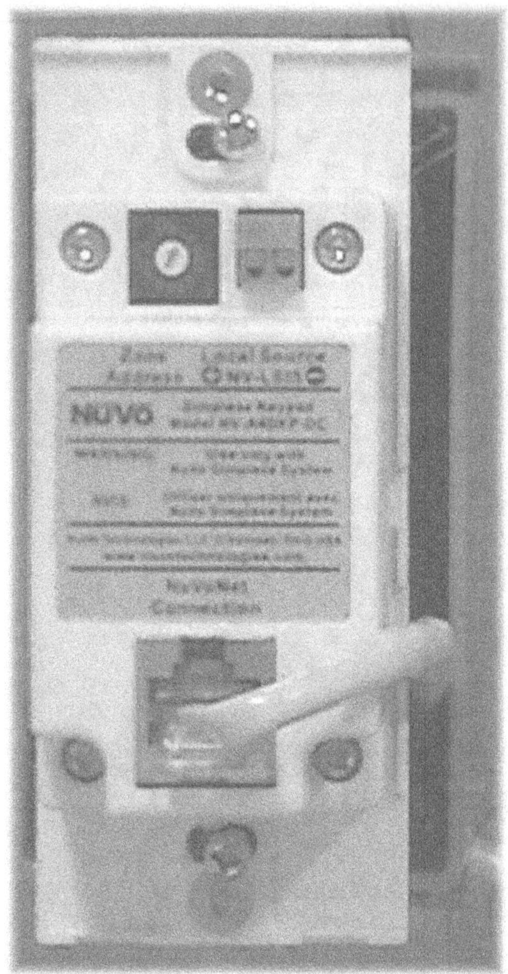

Figure 2-7: Connecting the Zone 1 Keypad to the CAT5 RJ-45 Cable

5. Secure the **Zone 1** keypad to the single-gang plastic junction box using the Phenolic 1 gang decorator cover plate and screws.

6. Complete Table 2-3.

7. Set the rotary switches on the keypads for Zones 2−4 to match the configuration recorded in Table 2-3.

8. Connect the CAT5 RJ-45 connectors from the single-gang plastic boxes for Zones 2−4 to their respective keypads.

Table 2-3: Zones 2-4 Keypad Rotary Switch Settings

Zone 2 Keypad Rotary Switch Settings			
Switch	**1**	**2**	**3**	**4**
Set				
Zone 3 Keypad Rotary Switch Settings			
Switch	**1**	**2**	**3**	**4**
Set				
Zone 4 Keypad Rotary Switch Settings			
Switch	**1**	**2**	**3**	**4**
Set				

9. Secure the keypads for Zones 2-3 to their respective single-gang plastic junction boxes, using the Phenolic 1 gang decorator cover plate and screws.

10. Connect the four **CAT5 RJ-45** terminated cables in the single-gang plastic box under the flat screen TV to the **AllPort Connection Hub**. See Figure 2-8.

NOTE: Connection arrangement does not matter.

11. Secure the **AllPort Connection Hub** assembly to the single-gang plastic box using the screws supplied with the unit.

12. Connect one end of the **AllPort Cable** to the front port of the **AllPort Connection Hub**.

13. Connect the remaining end of the **AllPort Cable** to the **AllPort Connection** of the dual-source, four-zone audio distribution system. See Figure 2-9.

Figure 2-8: Connecting the Keypads to the AllPort Connection Hub Assembly

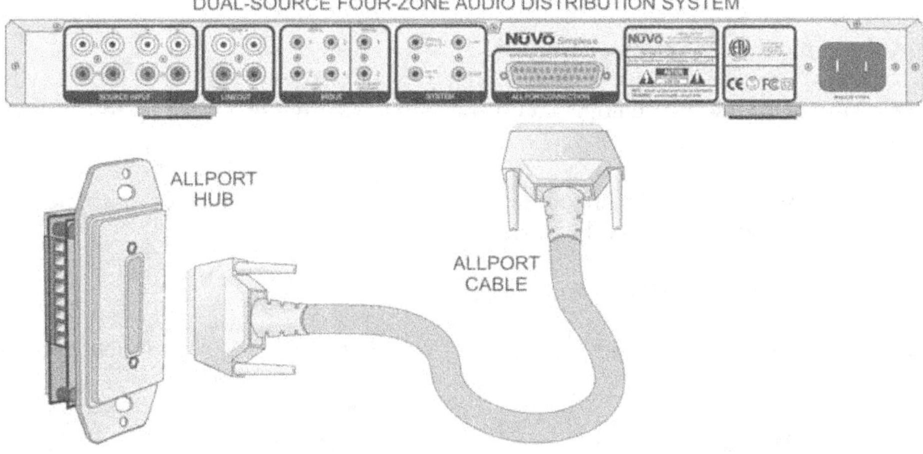

Figure 2-9: Connecting the Dual-Source, Four-Zone Audio Distribution System to the AllPort Connection Hub

14. Use the tie-wraps on the wiring at the audio distribution panel to keep all of the cables neatly organized.

Testing the Keypads and Zone Speakers

1. Review the section in the manufacturers manual pertaining to keypad usage.

2. Supply power to all of the devices.

3. Turn on all of the equipment by pressing the **power** buttons on all of the devices.

4. Put the **Avia Guide to Home Theater DVD** into the DVD player (**Source 2**).

NOTE: If any of the four zone speakers are on at this point, press the OFF button found on each zone keypad. See Figure 2-10.

Figure 2-10:
Dual-Source,
Four-Zone Audio
Distribution System
Keypad

5. Ask the instructor for the proper AM/FM station to tune into for the local area.

   Record the AM/FM radio station number: _________________.

6. Press the **source** button on the digital home audio and video entertainment system remote control (Source 1).

7. If necessary press the **source** button again on the digital home audio and video entertainment system remote control to switch between the AM and FM bands.

8. Set the sound level for the **AM/FM** tuner to **no more than 16**.

NOTE: Do not proceed until an audible sound is produced through the surround sound speakers from the AM/FM tuner.

9. Mute the five surround sound speakers by pressing the **mute** button on the remote control. The button might appear as an icon, a speaker that is X-ed out.

10. Press the **play** button on the remote control for the DVD player (**Source 2**).

11. Starting with the **Zone 1** keypad (see Figure 2-10), press the **ON/OFF** button once to turn the Zone 1 speaker **on**.

12. Press the **1** and **2** buttons to switch between the "Source 1" and "Source 2" hardware.

13. Check to see that the **volume** buttons are functioning by increasing and decreasing the volume slightly.

14. Press the **OFF** button on the Zone 1 keypad.

15. Repeat steps 10–14 for the remaining three zones.

Testing the IR Emitters

1. Turn on the **Zone 1** keypad.

2. Press the "**2**" button to select **Source 2**.

3. Hold the DVD player remote control in front of the **Zone 1** keypad.

4. Press the **OPEN/CLOSE** button on the remote control.

5. Watch the IR emitter on the front of the DVD player.

NOTE: It should flicker red and the DVD drawer should open.

6. Remove the **Avia DVD** from the DVD player (Source 2) and return it to its jacket.

7. Hold the DVD player remote control in front of the **Zone 1** keypad and press the **OPEN/CLOSE** button a second time to close the drawer.

8. Hold the DVD player remote control in front of the **Zone 1** keypad and press the **Power** button to turn **off** the DVD player unit.

NOTE: The DVD player should power down.

9. Press the "**1**" button on the **Zone 1** keypad to switch to the **Source 1** digital home audio and video entertainment system.

10. Hold the digital home audio and video entertainment system's remote control in front of the **Zone 1** keypad.

11. Press the **Power** button on the digital home audio and video entertainment system remote control and turn the unit off.

NOTE: The digital home audio and video entertainment system should display the goodbye message and go into standby mode.

12. Turn off all of the keypads.

13. Turn off the dual-source, four-zone audio distribution system.

14. Turn off the television monitor.

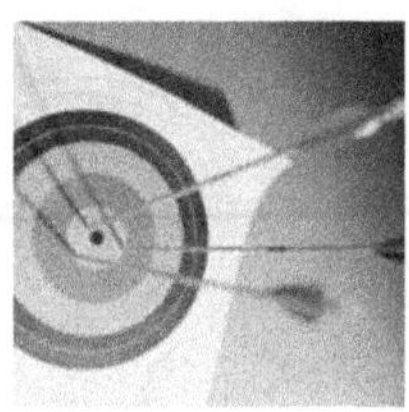

Feedback

LAB QUESTIONS

1. Where should the audio distribution source equipment be located?

2. What is a zone?

3. What are the two types of distribution plans for audio sources?

4. What equipment might be included in a basic audio source equipment list for a zoned system?

5. What does a single-zone system do?

Adjusting LCD TV Monitors and Surround Sound Audio

OBJECTIVES

1. Adjust the LCD TV monitor's brightness, color, contrast, tint, sharpness, and backlight to the NTSE standard.
2. Perform the Avia DVD monitor adjustment procedures.
3. Test the different adjustment settings provided on the equipment for use with the surround sound speakers.

Audio/Video Fundamentals

RESOURCES

1. Marcraft Audio/Video Experiment Panel and Frame
2. Dual-source, four-zone audio distribution system
3. DVD player system
4. Digital home audio and video entertainment system
5. LCD TV monitor
6. Avia Guide to Home Theater DVD with color filters (red, green, and blue)

DISCUSSION

Liquid Crystal Displays

Liquid Crystal Displays (LCDs) are the most common type of flat panel display. They are used for a wide range of small portable electronic and computing devices including laptop computers, cell phones, games, and Personal Digital Assistants (PDAs). LCD displays utilize two sheets of polarizing material with a liquid crystal solution between them. An electric current passed through the liquid aligns the crystals so that light cannot pass through them. Each crystal, therefore, is like a shutter, either allowing light to pass through or blocking the light.

Color LCD displays are created by adding a three-color filter to the panel. Each pixel in the display corresponds to a red, blue, or green dot on the filter. Activating a pixel behind a blue dot on the filter will produce a blue dot onscreen. Like color CRT displays, the dot color on the screen of the color LCD panel is established by controlling the relative intensities of a three-dot (RGB) pixel cluster.

A display screen made with Thin-Film Transistor (TFT) technology is an advanced liquid crystal display design, common in top-of-the-line laptop computers, that has a transistor for each pixel. TFT is also known as "active matrix" display technology and contrasts with "passive matrix" LCD technology, which does not have a transistor at each pixel. An active matrix display can be switched on and off very rapidly, presenting a fast, high-resolution appearance. Active matrix is the superior technology for LCD displays.

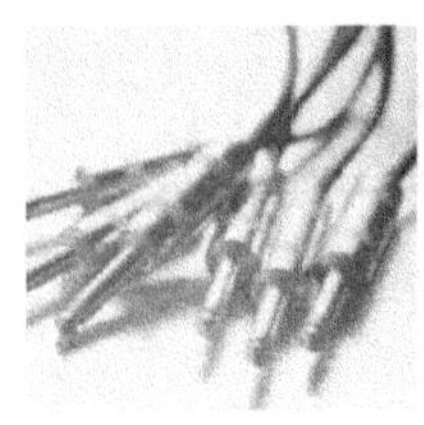

PROCEDURES

Adjusting the Display

NOTE: Confer with your instructor for proper setup before performing this procedure.

1. Review the section of the manufacturer's manual for the LCD TV monitor that pertains to the TV picture menu controls.

2. Put the **Avia DVD** into the DVD player.

3. Use the remote control for the DVD player to navigate to the **Main Menu**.

4. Select **Audio & Video Calibrations**.

5. Select **Basic Video Adjustments**.

6. **Rewind** the video segment and **review** the information as many times as necessary before performing the adjustments.

7. Perform the adjustments as they are called for during the video segments.

NOTE: If the display adjustment menu obstructs your view while making any of the following adjustments, make an adjustment and close the menu. Recheck the pattern and make additional adjustments as necessary.

White Level (Contrast or Picture)

Power supply adequacy, found in CRT units, causes line bending when the white level is adjusted too high. Watch for the beginning of line bending but don't adjust the white level to the point of blooming. Blooming is caused when the display loses focus of its' electron beam. Blooming and power supply adequacy, as they pertain to CRT monitors, are covered fully in the video segment on white level found on the Avia DVD.

Thresholding occurs in LCD projectors if the white level is too high. Highlight details are lost in the white background. If one or two of the white bars in the test pattern disappear into the white background, then white thresholding is occurring. Remember the setting used when white thresholding occurs as this is the maximum setting for white level on the display.

NOTE: NTSE standards — The Avia DVD video will attempt to bring the display device as close to NTSE standards as possible. These are the standards to which reference monitors are tuned when used during the DVD creation process.

8. Set the room lighting at levels appropriate for viewing conditions.

9. Use the remote control for the LCD TV monitor to make these adjustments.

10. Press the **Menu** button on the LCD TV monitor remote to bring up the adjustment menu on the monitor.

11. Wait for the Avia DVD video to display the needle pulses and steps pattern.

12. Set the white level so that the bottom part of the pattern appears gray.

13. Gradually increase **white level** until the bottom portion of the pattern appears white.

Black Level (Brightness)

Black level is used to set the appearance of black and the transparency of an image. If the black level is set too high, the pictures may seem washed out. If it is set too low, shadow details will be lost into black. When black level is set correctly the black background of the test pattern is completely black, the leftmost moving bar is essentially invisible, and the rightmost moving bar is just visible.

14. Using the black bar and half gray test pattern, adjust the **black level** so that the left moving bar is essentially invisible and the right bar is just barely visible.

15. Recheck white level if necessary.

Sharpness (Peaking, Detail)

Sharpness brings out detail without adding extraneous outlines or noise to an image. Sharpness when set correctly will preserve details and improve the picture fidelity. If the sharpness is set too low the image will appear blurred.

16. Turn color saturation all the way down, making the display as black-and-white as possible.

17. Adjust **sharpness** up and down on the display.

18. Watch the brightness of lines in the frequency sweep at the top of the screen.

19. Also observe the patches of lines at the bottom of the screen.

NOTE: Some of the patches and a section of the frequency sweep at the top of the display will grow brighter as sharpness is increased. This is caused by the control emphasizing those video frequencies. When sharpness is correctly set, all lines of the sweep will appear equally bright. The patches at the bottom of the pattern are equally bright, except for the rightmost patch, which is often slightly darker.

20. Observe the lines and the circle in the center of the pattern while adjusting sharpness. When the control is set too high, false white outlines will appear next to the black lines. When set too low, the black lines become blurred.

21. Adjust the **sharpness** control to make the false outlines disappear, the line patches equal in brightness and the frequency sweep evenly bright.

Saturation (Color, Chroma) and Hue (Tint)

22. Check to determine whether the display has automatic flesh tone controls. If it does, then turn this control setting **off**. This setting makes it almost impossible to achieve NTSE accurate color.

NOTE: The display we are using doesn't have the automatic flesh tone control feature.

NOTE ON SATURATION: Color bar patterns are used for adjusting color controls. When saturation and hue are correctly set, the gray, blue, cyan, and magenta portions of this pattern all contain the same amount of blue. We will achieve the correct results by viewing the display through a blue filter. A display's saturation control adjusts the amount of color in its picture. Pay attention to the left and right bars and the patches in the test pattern while adjusting the saturation setting. The pattern must be viewed through a blue filter while making adjustments. Adjust saturation until flashing of the saturation patches is minimized. The gray and blue bars and patches of the pattern will also become equal in blue brightness when saturation is correct.

NOTE ON HUE: The hue control adjusts the tint of colors. Again watch the display through a blue filter. Pay attention to the central hue portion of this pattern while adjusting hue. Adjust until flashing of the hue patches is minimized and the magenta and cyan areas are equal in brightness.

23. Use the blue filter to adjust the **saturation** patches so that flashing is minimized.

24. Use the blue filter to adjust the **hue** patches so that flashing is minimized.

NOTE: If the display adjustment menu obstructs your view while making saturation and hue adjustments, make an adjustment and close the menu. Recheck the pattern through the filter and make additional adjustments as necessary.

25. Use the **three colored filters** and the **color decoder check pattern** to complete Table 3-1.

Table 3-1: Color Emphasis Percentage from Decoder Check Pattern and Filter

	Red	Green	Blue
Percentage (+ or -)			

26. Remove the **Avia DVD** from the DVD player (Source 2) and return it to its jacket.

27. Turn **off** all the equipment.

Testing the Surround Sound Speakers

1. Power up the digital home audio and video entertainment system and the 15-inch flat screen TV.

2. Review the active navigation section from the manufacturer's instruction manual for the digital home audio and video entertainment system.

3. Insert the **Avia Guide to Home Theater DVD** into the DVD player.

4. Adjust the **volume** as necessary.

5. When the DVD starts to play, press the **MENU** button on the remote control for the digital home audio and video entertainment system.

6. Press the **down arrow** button until the *Advanced AVIA complete audio and video calibrations* option is highlighted for selection.

7. Press the **OK** button.

8. Press the **down arrow** button until the *Verification/Evaluation* option is highlighted.

9. Press the **OK** button twice.

10. Verify that each speaker is connected and functioning correctly as the tests are performed by putting a check mark in the corresponding column of Table 3-2 as each test is completed.

Table 3-2: Surround Sound Audio Test Results

	LF		C		RF		RS		LS	
	Heard	Not Heard	Heard	Not Heard	Heard	Not Heard	Heard	Not Heard	Heard	Not Heard
Wideband Pink, 5 Channel Pan										
I 50 HighPass Pink, 5 Channel Pan										
Circulating Ambience Generator Clicks										
Pink Noise Match of Center Speaker										
Low-Frequency Sweep, Left-Front										
Low-Frequency Sweep, Center										
Low-Frequency Sweep, Right-Front										
Low-Frequency Sweep, Right-Surround										
Low-Frequency Sweep, Left-Surround										

11. Remove the **DVD** from the DVD player and return it to its jacket.

12. **Power down** both devices.

Testing Surround Sound Output Modes

1. Turn **on** the digital home audio and video entertainment system.

2. Ask the instructor for the proper local **AM/FM station** to tune the receiver to.

    Record the AM/FM radio station number: _________________.

3. Press the **tuner** button on the digital home audio and video entertainment system remote control (Source 1).

4. If necessary, press the **tuner** button again on the digital home audio and video entertainment system remote control to switch between the AM and FM bands.

5. Set the sound level for the AM/FM tuner to **no more than 16**.

6. Review the section on selecting surround output modes in the instruction manual for the digital home audio and video entertainment system.

7. Use the **remote control** for the digital home audio and video entertainment system to run through the different surround output modes.

8. Complete Table 3-3. Refer to Figure 3-1 for the location of the speaker indicator in the digital home audio and video entertainment system.

Figure 3-1: Location of the Surround Sound Speaker Display Indicator

Table 3-3: Speakers Affected by Surround Output Modes

	LF		C		RF		RS		LS		SUB	
	On	Off	On	Off	On	Off	On	Off	On	Off	On	Off
Prologic												
3-Stereo												
Stereo												
Stadium												
Stage												
Hall												
Disco												
Live												

9. Turn **off** the digital home audio and video entertainment system.

LAB QUESTIONS

1. How does an LCD display work?

2. What is the most common type of flat panel display?

3. What type of technology is used to create an advanced liquid crystal display?

4. What is the function of the black level control?

5. What does the sharpness control do?

6. What happens when saturation and hue controls are adjusted correctly using a test pattern?

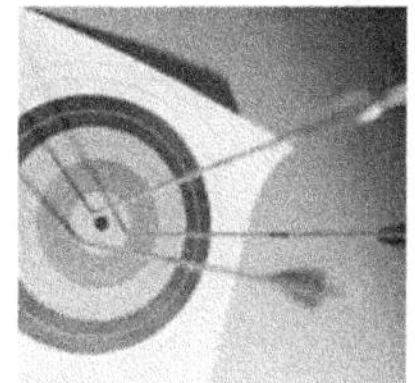

Feedback

Audio and Video Fundamentals Research

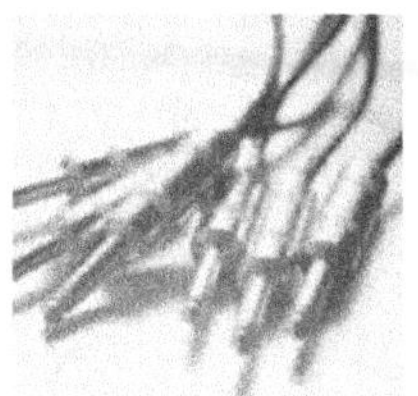

**Audio/Video
Fundamentals**

OBJECTIVES

1. Search for external audio and video services.
2. Research various audio and video media servers and components.
3. Locate applicable local, state, and federal regulations.

RESOURCES

1. Books
2. Newspapers
3. Magazines
4. Computer with graphic capture software and Internet access

DISCUSSION

One of the most underrated skills required for successful scholarship is the ability to conduct research. Numerous sources of information are available to the modern student, so the task of gathering meaningful information is not as difficult as it once was. While traditional researching methods continue to require a fair amount of card file searching and reading (books, newspapers, magazines), the Internet has opened up new opportunities for conducting information searches about almost any subject of interest. Modern libraries are equipped with computer terminals for conducting Internet searches.

In fact, by searching the Internet without basic researching skills, inquisitive minds often run the risk of drowning in an ocean of unwanted information. The trick is to locate the desired information quickly, without becoming ensnared in time-wasting searches through mountains of available data. Online information items are not individually catalogued for identification and retrieval. Billions of individual files are being added to the Internet even as you read this, and many data delivery services are available from which to choose. Among these are electronic mail, file transfers, interest group memberships, interactive collaborations, and multimedia displays.

The World Wide Web (WWW) includes most of the Internet protocols required to effectively handle the data services previously mentioned. But, although the Web offers access to much of what is available on the Internet, the addresses of Internet sites frequently change, or disappear altogether. When this happens, screens similar to the one shown in Figure 4-1 often appear.

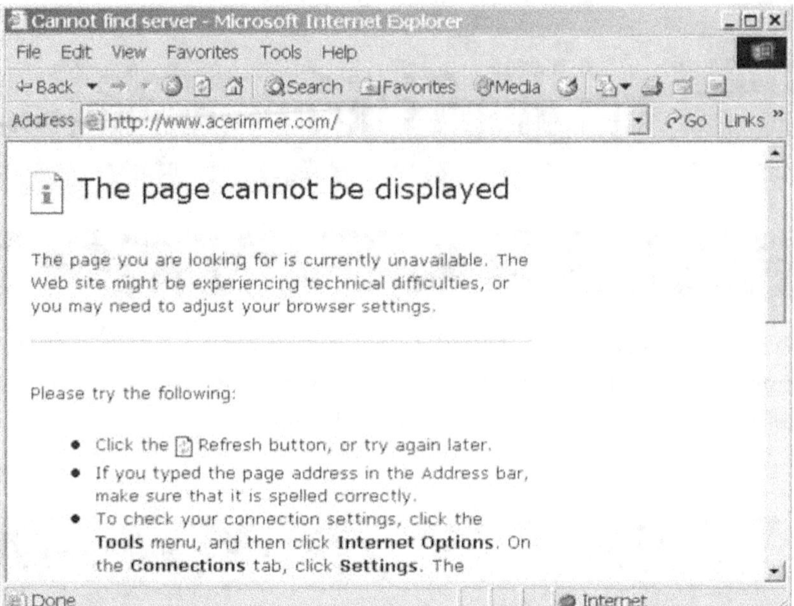

Figure 4-1: Internet Site or Page Cannot be Displayed

Do not expect stability on the Internet! An efficient method of conducting Internet research is to copy and store any information (text and graphics) that appears useful to the project at hand. If later on, the Internet site from which the material was discovered can no longer be found, the information itself is still available to you. In addition, certain Internet sites contain information that is impossible to transfer directly to your computer. This situation usually occurs when a product vendor has provided proprietary information. In these cases, it may be necessary to type the relevant text, by hand, to a word processor for data storage. Graphics or pictures may have to be captured using a third-party graphics tool, and then converted into a usable picture format.

One other important consideration for conducting research, especially on the Internet, is to verify the integrity of the information you use. Although someone has placed some interesting tidbits on an Internet site, or in a newspaper, magazine, or book article, this does not guarantee their accuracy or truthfulness. Identical information from a variety of sources tends to strengthen the case for veracity!

PROCEDURES

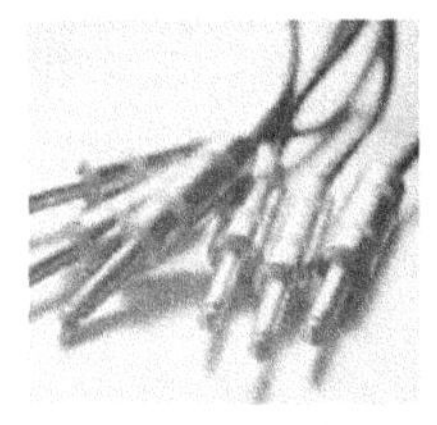

Audio/Video Fundamentals

Because external audio and video services are a vital concern to total connectivity, your beginning research should be concerned with identifying those services and understanding how they augment the usability of the personal computer.

1. Examine the following list of external audio and video services and components. The RESI Audio and Video technician will need to be familiar with the items shown here. A codec is a compressor/decompressor, which can be either a software or hardware digital data manipulator.

2. Use Tables 4-1 through 4-8 to organize the specified details about the external audio/video services and components listed above. For each item, try to locate at least two vendors.

- Audio/video storage
- External soundboards
- Hardware encoders
- Internet broadcasting

- Network bandwidth
- Servers
- Software encoders
- Windows A/V codecs

Table 4-1: Audio/Video Storage

Product	Description	Vendor	Cost

Table 4-2: External Soundboards

Product	Description	Vendor	Cost

Table 4-3: Hardware Encoders

Product	Description	Vendor	Cost

Table 4-4: Internet Broadcasting

Product	Description	Vendor	Cost

Table 4-5: Network Bandwidth

Product	Description	Vendor	Cost

Table 4-6: Servers

Product	Description	Vendor	Cost

Table 4-7: Software Encoders

Product	Description	Vendor	Cost

Table 4-8: Windows A/V Codecs

Product	Description	Vendor	Cost

Another concern for the RESI Audio and Video technician is the various components that make up a media server. Time constraints and the need for media sharing make the use of a media server more and more imperative.

So as to make your remaining research time more productive, concentrate on information about the server components in the product areas listed below. This researching task will be of benefit when considering the options for a media server installation, including its size, layout, and the particular requirements of the client. Information gathered here will help when considering and comparing specific pieces of server equipment, their capabilities, available vendors, and estimated costs.

3. Examine the following list. It identifies media servers and software with which a RESI Audio and Video technician will most likely be working.

4. Use Tables 4-9 through 4-28 to organize the specified details about the media servers and components listed above. For each item, try to locate at least two vendors.

- 2NetFX servers/players
- BroadWare streaming servers
- Burstware media servers
- Cashflow servers
- Content delivery networks
- DBS streaming software
- GMV encoders/players
- Icecast streaming audio
- ImageOn java decoder
- iVast servers/players

- LessData video player
- Mediabase VCD
- Personal video recorders
- Philips digital networks
- Streaming21 servers/players
- QuickTime streaming servers
- RealPlayers
- Universal servers
- Virage video servers
- Windows servers

Table 4-9: 2NetFX Servers/Players

Product	Description	Vendor	Cost

Table 4-10: BroadWare Streaming Servers

Product	Description	Vendor	Cost

Table 4-11: Burstware Media Servers

Product	Description	Vendor	Cost

Table 4-12: Cashflow Servers

Product	Description	Vendor	Cost

Table 4-13: Content Delivery Networks

Product	Description	Vendor	Cost

Table 4-14: DBS Streaming Software

Product	Description	Vendor	Cost

Table 4-15: GMV Encoders/Players

Product	Description	Vendor	Cost

Table 4-16: Icecast Streaming Audio

Product	Description	Vendor	Cost

Table 4-17: ImageOn Java Decoder

Product	Description	Vendor	Cost

Table 4-18: iVast Servers/Players

Product	Description	Vendor	Cost

Table 4-19: LessData Video Player

Product	Description	Vendor	Cost

Table 4-20: Mediabase VCD

Product	Description	Vendor	Cost

Table 4-21: Personal Video Recorders

Product	Description	Vendor	Cost

Table 4-22: Philips Digital Networks

Product	Description	Vendor	Cost

Table 4-23: Streaming21 Servers/Players

Product	Description	Vendor	Cost

Table 4-24: QuickTime Streaming Servers

Product	Description	Vendor	Cost

Table 4-25: RealPlayers

Product	Description	Vendor	Cost

Table 4-26: Universal Servers

Product	Description	Vendor	Cost

Table 4-27: Virage Video Servers

Product	Description	Vendor	Cost

Table 4-28: Windows Servers

Product	Description	Vendor	Cost

Applicable Regulations

A completed audio and video system installation will testify either to the skill and awareness of the RESI Audio and Video technician who installed it, or to his/her failure to consult the local, state, or federal regulating bodies having jurisdiction over residential or commercial audio/video systems.

Local

While strict adherence to federal and state regulations will often result in totally legal audio and video equipment installations, this does not guarantee an automatic free pass by the local coding authority. Special local ordinances must be taken into account before any A/V system installation begins! These codes are likely to involve local sound ordinances, rather than regulate video installations. However, attempts to control access to specific types of undesired content may have resulted in certain city or county restrictions being activated. Some local governments may have codes that are designed to protect community standards regarding audio and video content and include certain civil penalties for chronic violators. Many local enforcement agencies will take action if the community is exposed to loud or obscene audio or video content, including subwoofer operations that can be heard or felt throughout a neighborhood, or large-screen TV presentations of pornographic material that may be publicly viewed from adjacent properties.

Information concerning applicable community standards for A/V installations can be gathered from local planning commissions or coding departments. These bodies are usually responsible for setting and enforcing the standards of decency in the local community. Try contacting your local planning commission and gather any specific information about requirements for audio and video equipment not specifically addressed by state or federal codes. Such guidelines could be enforced by either city or county agencies. For example, certain citizens groups may request that unique restrictions be placed on the local access of certain types of audio or video media and/or equipment to protect children from exposure to them.

1. Use Table 4-29 to organize specific details about media server installations and/or equipment strictly controlled by your local codes. Try to locate more than one source of information for each consideration or restriction.

Table 4-29: Local Media Server Codes or Regulations

Information Source	Specified Server Installation	Restrictions	Special Considerations

State

State-sponsored media regulations include the protection of copyrights, including the downloading or duplication of audio, video, or gaming materials from state governmental or educational servers. State agencies and universities regulate the legitimate use of their computers. Their policies normally restrict usage to conducting legitimate government or college business, and supporting various departmental mandates or institutional missions. These policy guidelines extend to systems that lie outside state or college networks, but which are nevertheless accessed state-sponsored facilities. Computing or network providers outside of state or university jurisdiction often impose additional conditions for appropriate use, for which users at the state or university level are responsible. As might be expected, these conditions also apply to outside networks used to access state governmental or university networks.

Most state governmental and educational jurisdictions confine the use of their computer and network resources to official legislative and administrative functions. Agreement generally exists within these agencies that such resources must not be used for personal, commercial, or for-profit purposes without some form of official written approval. As might be expected, approvals of this type are extremely rare.

Consistent with most state governmental and educational policies on equal opportunity and affirmative action, state-sponsored computer and network resources may not be used to store, transmit, or receive any text, image, audio, or video materials that are discriminatory, abusive, profane, threatening, harassing, or sexually offensive.

Infringement of copyright laws and the appearance of obscene, harassing, or threatening material on state governmental or educational servers usually violate local, state, national, or international laws. Therefore, such actions are subject to litigation by the appropriate law enforcement agency. Such has been the experience of college students recently sued by the Recording Industry Association of America (RIAA) for operating small on-campus file-sharing networks. Various campus communities have begun to address this problem, warning of the possible consequences for those individuals using file-sharing software illegally. Media industries have been eager to stop their unprecedented sales decline, which they see as due in part to the ease with which files are downloaded and burned onto blank CDs or DVDs. In order to discourage the appearance of inappropriate or copyrighted material, most state governmental and educational network policies also prohibit their telephone systems from being used to access the Internet.

Individuals residing on networks outside of state jurisdiction, including residential networks, can also be prosecuted for any of the above infractions originating from their computers. Information concerning the legitimate usage of state governmental and educational media servers in your state can be found by browsing government Web pages.

2. Search the information pages of several universities in your state and locate the applicable server policies regarding the observance of audio copyrights. Organize their specific details in Table 4-30.

Table 4-30: State University Audio Media Server Policy

University or Agency	Specified Copyrights	Distribution Restrictions	Exceptions Permitted

3. Search the information pages of several universities in your state and locate the applicable server policies regarding the observance of video copyrights. Organize their specific details in Table 4-31.

Table 4-31: State University Video Media Server Policy

University or Agency	Specified Copyrights	Distribution Restrictions	Exceptions Permitted

4. Search the information pages of several universities in your state and locate the applicable server policies regarding the observance of gaming copyrights. Organize their specific details in Table 4-32.

Federal

Table 4-32: State University Gaming Media Server Policy

University or Agency	Specified Copyrights	Distribution Restrictions	Exceptions Permitted

Various technical bodies generally recognized by manufacturers and consumers around the world have the authority to determine the suitability of one product over another. The explosion of technical development during the last several years is continuing, and in order to attain some control over this activity, the production of suitable standards has been deemed necessary.

The major organizations listed here publish standards that determine the acceptability of both products and labor involved in computer media server system installations and operations. Because they determine the suitability of one product over another, the RESI Audio and Video technician can use the standards published by these organizations to select appropriate equipment for a particular installation.

National Electrical Code

Sponsored by the National Fire Protection Agency, the National Electrical Code contains the information upon which all aspiring electricians must be tested before obtaining their licenses. Having spent so much time studying it, most electricians can easily locate any required information in the code. Other users should use the index. The National Electrical Code is merely a guideline, and most states require a permit and an inspection for performing electrical work. Following the NEC does not guarantee safe electrical installations, but it is the best guide available. Given the many applications available for specific wiring jobs, an electrician should use his or her own judgment, while keeping above the minimum safety standards set forth by the NEC.

Because each state may differ slightly in its requirements for inspection and code compliance, the local city or town wire inspector should be contacted before wiring any media server. The local wire inspector is "The Authority Having Jurisdiction" in most locations, and as such is responsible for rule interpretation and for code enforcement.

5. Identify the local wire inspector in your area. Organize the details about your local wire inspector in Table 4-33.

Table 4-33: The Authority Having Jurisdiction

Business Name	Personal Name	Street Address	Telephone Number	Web Address

TIA/EIA Standards

The Telecommunications Industry Association (TIA) and Electronic Industries Alliance (EIA) standards are certified by the American National Standards Institute (ANSI) while Telecommunications Systems Bulletins (TSBs) are addenda to, or explanatory comments about, either an industry standard or an interim standard, and are often integrated into the next standard revision.

6. Conduct an Internet search for any ANSI/TIA/EIA standards pertinent to residential media server operations, and write information about them in Table 4-34.

Table 4-34: ANSI/TIA/EIA Residential Media Server Operations

Standard ID	Standard Name	Standard Description	Year Established

IEEE Standards

The IEEE, the Institute of Electrical and Electronics Engineers, Inc., a nonprofit, technical professional association of more than 377,000 individual members from more than 150 different countries. Because of the size and scope of its membership, the IEEE has become a leading authority in technical areas ranging from computer engineering, biomedical technology, telecommunications, electric power, aerospace, and consumer electronics.

7. Conduct an Internet search for any IEEE standards pertinent to residential media server operations, and write information about them in Table 4-35.

Table 4-35: IEEE Residential Media Server Operations

Standard ID	Standard Name	Standard Description	Year Established

Underwriters Laboratories Inc. (UL)

As an independent, not-for-profit, product safety testing and certification organization, Underwriters Laboratories (UL) has tested products for public safety for more than 100 years. Every year, more than 17 billion UL marks are applied to products from all over the world. Founded in 1894, it is the undisputed leader in U.S. product safety and certification. But, the UL is becoming one of the most recognized, reputable conformity assessment providers in the world, not just the United States.

At the time of this publishing, Underwriters Laboratories help companies achieve global acceptance for their product, whether it is an electrical device, a programmable system, or a company's quality process. UL lists various manufacturers for specific types of installations and equipment, tests the equipment for specific uses, and judges installations against the governing standards.

8. Using an Internet search, locate several UL standards that manufacturers have incorporated into some of the computer media server components listed earlier. Include the information you find in Table 4-36. Feel free to extend the table if necessary.

Table 4-36: UL Computer Media Server Standards Incorporated by Manufacturers

Server Component	Manufacturer	Standard ID	Specified Requirements

National Fire Protection Association (NFPA)

In an effort to reduce the worldwide burden of fire and other hazards on quality of life, the National Fire Protection Association (NFPA) has developed and advocated a scientifically based consensus of codes, standards, research, training, and education since its founding in 1896. Today, the NFPA, with its headquarters located in Quincy, Massachusetts, is an international, nonprofit membership organization
that supports more than 75,000 members representing nearly 100 nations, and 320 employees around the world. In addition to setting installation guidelines for fire equipment, the NFPA also is the world's leading advocate of fire prevention. It holds classes to instruct various installers on how to follow the guidelines, and is an authoritative public safety source.

9. Using an Internet search, locate several NFPA standards that manufacturers have been incorporated into some of the computer media server components listed earlier. Include the information you find in Table 4-37. Feel free to extend the table if necessary.

Table 4-37: NFPA Computer Media Server Standards Incorporated by Manufacturers

Server Component	Manufacturer	Standard ID	Specified Requirements

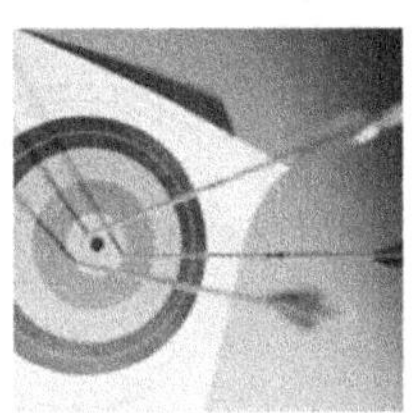

Feedback

LAB QUESTIONS

1. List several popular formats for audio and video file storage.

2. What do media industries consider to be part of the cause for their unprecedented sales decline?

3. Who certifies the TIA/EIA standards?

4. Why can't most electricians easily locate any required information in the massive NEC?

5. What does the acronym RIAA stand for?

MultiRoom Video Distribution Systems

OBJECTIVES

1. Identify the home theater requirements for the given scenario.
2. Identify the video display technologies needed to meet the customer's requirements.
3. Identify the cabling infrastructure and components that are required to distribute satellite and off-air TV services throughout the house.
4. Identify a common interference problem that occurs during home theater installation.
5. Recommend appropriate video displays for rooms that have different and unique requirements.
6. List the features that need to be identified when purchasing the new display units.
7. Recommend a suitable cabling infrastructure.
8. List some considerations when installing the LCD in the kitchen.
9. Troubleshoot an interference problem.
10. Choose a cable to interconnect the digital satellite set-top box with the front projection system.
11. Define a mechanism for programming remote control buttons.

Audio/Video Fundamentals

RESOURCES

1. Marcraft's *RESI Audio/Video Systems Endorsement* textbook
2. Pencil

DISCUSSION

Read the following scenario:

Founded in 1999, Theater Technologies Ltd. is a systems installation firm that provides custom electronic installation services to the New York area. The company designs and installs custom home theaters and multiroom music systems uniquely tailored to each client's home, lifestyle, and budget. You are a system consultant for Theater Technologies Ltd. and you have been recently assigned to a number of different projects.

PROCEDURES

Home Theater Technologies

The first project involves the design of an integrated distributed video system for a new residence. The house is approximately 6,000 square feet in size, and is currently under construction.

Your task is to help the homeowner decide on a system that best fits their needs. This is achieved by completing the following steps:

- Recommend a display unit.
- Identify the impact of installing a video projector unit.
- Complete the equipment list for the display unit.

1. The client wants to use a display unit to produce images spanning several feet across the room. Which of the following video display technologies will satisfy the homeowner's viewing taste in the home theater room?

    ____a. Video projection (a front projection system)
    ____b. Flat-panel
    ____c. Cathode ray tube

Why?

____

____

2. What impact will a video projector in the home theater have on the project?

____

____

____

3. The products required for the home theater room include:

- a big video screen
- five speakers (front left, center, front right, back left, and back right)
- material for the room
- an integrated lighting system
- some surge suppressors
- data and cable wall plates
- a control device
- specialty home theater seating

What equipment do you think has been omitted from the equipment list?

____

____

Display Technologies

Your second housing project involves the installation of several different types of display technologies.

1. Complete Table 5-1 by recommending the appropriate video displays for rooms that have different and unique requirements.

Table 5-1: Recommended Video Displays

Room and Requirements	Suggested Type of Video Display	Customers Viewing Taste
Home theater		A display that spans several feet across the room and accepts multiple sources ranging from VCR to computers
Kitchen		A flat-panel display that produces flicker-free images in a small space
Main bedroom		A display that will allow use of a keypad
Children's room		A relatively low-cost 20-inch TV that displays a bright picture with high contrast

2. A few key features can dramatically affect the performance of the display unit. Use Table 5-2 to map the functionality required by the homeowner for the new home theater display unit by providing a list of technical requirements that may be referenced during the purchase of the new unit.

Table 5-2: Functionality and Technical Requirements

Homeowner Required Functionality	Technical Requirements
Support for HDTV	
Two digital connectors	
A high-quality analog connector	
Wide screen support	

Video Distribution

The third project for Theater Technologies Ltd. involves the distribution of satellite and terrestrial TV signals throughout the home. The project will install a satellite digital TV system with two set-top boxes, one in the home theater (family) room and the other in the living room, to allow family members to watch different channels. Additionally, the client wants to access terrestrial channels in various parts of the house. Finally, the client has requested the installation of a flat-screen LCD in the kitchen.

1. On Figure 5-1, draw and label your recommended cabling infrastructure and the components required for the acoustical system.

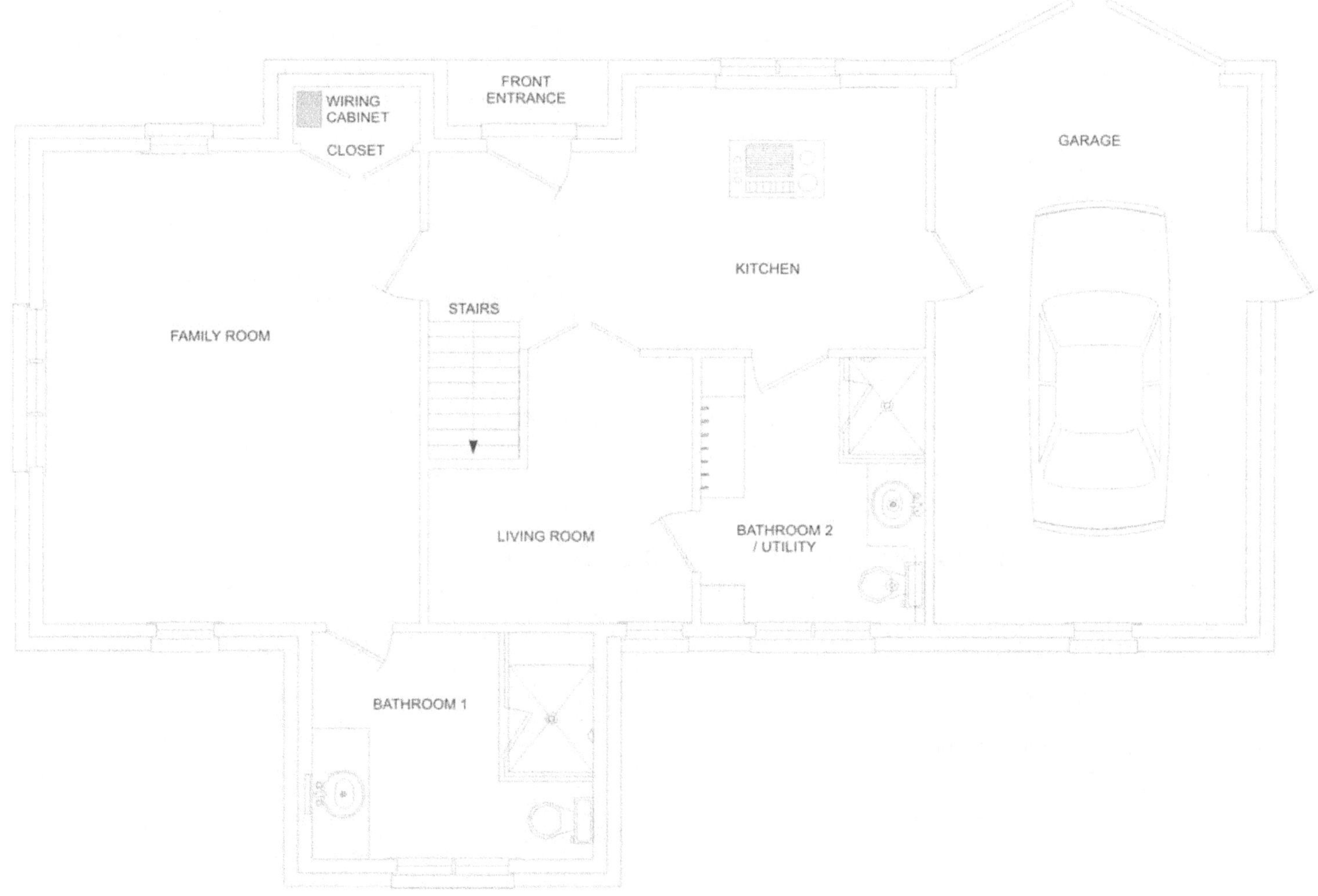

Figure 5-1: Your Recommended Cabling Infrastructure and Components

2. Which of the following precautions need to be taken when installing the flat-screen LCD in the kitchen?

    ____a. It needs to be shielded from bright sunlight and other heat sources such as the oven.
    ____b. It should be kept away from ultraviolet light sources.
    ____c. It should be kept away from the refrigerator to avoid exposure to cold conditions.
    ____d. It should be kept away from the seating area.

Why?

____

____

____

____

Configuration and Testing

Your final project involves testing a system.

1. While testing the home theater system in the previous project you notice some slight interference on the video display. You suspect that the interference is coming from the center speaker. Identify a solution that does not require the installer to relocate the speaker.

____

2. Which of the following cables should you use in the home theater to connect the digital satellite set-top box with the front projection system?

    ____a. A component video connection cable
    ____b. An S-Video cable
    ____c. A triple RCA cable arrangement

Why?

____

____

3. The homeowner wants to use one button on the remote control to turn on the TV, DVD player, and surround sound subsystems. What would you do to the home theater controller to fulfill this request?

____

____

____

Multiroom Audio Distribution Systems

OBJECTIVES

1. For the given scenario develop a plan for a new multiroom music system.
2. Identify and describe the physical audio products and cables that make up the new audio system.
3. Design a system to control audio sources from any room in the house.
4. Describe the purpose of the preamplifier.
5. Draw the cabling infrastructure for the new audio system.
6. Describe the type of amplifier required for the new system.
7. Identify the limitation of using a single amplifier.
8. Identify the cable type for this project.
9. Calculate the gauge of audio wire.
10. Recommend a speaker system for the home theater room.
11. Recommend a speaker system for the bedrooms and the bathrooms.
12. Define the ratings for the various speakers.
13. Suggest procedures for installing the speakers.
14. Configure the speaker system in the dining room.
15. Remotely control the audio sources.
16. Describe a keypad that consolidates control units.

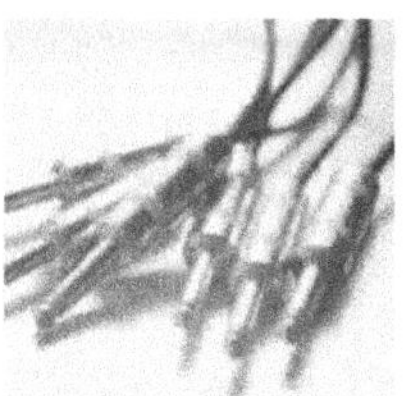

**Audio/Video
Fundamentals**

RESOURCES

1. Marcraft's *RESI Audio/Video Systems Endorsement* textbook
2. Pencil

DISCUSSION

Read the following scenario:

SmartProperties Ltd specializes in the custom design and installation of smart home systems. The company's services include programming and installation of products such as:

- Structured wiring for audio and video distribution
- Networking for data and phones
- HVAC
- Home theater
- Pool and irrigation
- Lighting control
- Home automation in general

The owner of a new home development — David Severn — has commissioned the company to design and install a multizone audio system. Table 6-1 includes specifications and client requirements gathered during interviews with David.

Table 6-1: Functionality and Technical Requirements

Requirement Category	Details
Management	The system will need to be managed at a central location.
Local control	The system should allow family members in each room to play music at a particular volume level.
Multisource	Different music sources should be available in different parts of the home. David plans to use four separate sources of music — a DVD player, a radio tuner, a digital audio server, and a CD player.
Reception	The tuner must be configured to receive a group of radio signals that are of the highest quality available off-air.
Home entertainment	The new audio system will need to support a dedicated home theater room.
Interior design	The speakers for the system should blend in with the décor of the house.
Background music	The ability to hear background music will be required in bedrooms.

PROCEDURES

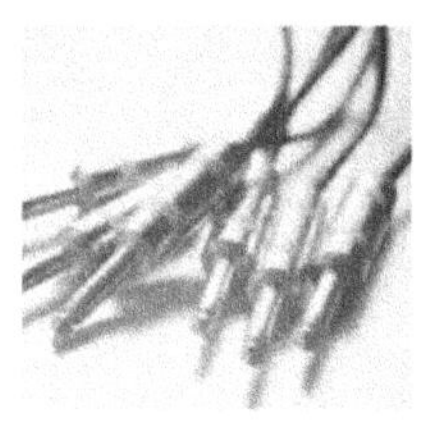

Audio/Video Fundamentals

Proposed Design Strategy

1. SmartProperties Ltd has subcontracted you as a system consultant to analyze a draft design (Table 6-2) compiled by external consultants, and to recommend adjustments.
    In light of the customer's requirements, was AM the best choice for receiving high-quality radio channels?

    ____a. Yes
    ____b. No

Table 6-2: Draft Design Plan

Design Element	Proposed Implementation
Type of audio system	Distributed
Tuner system configuration	AM
Equipment list	The proposed equipment list includes the following components: • Six pairs of speakers • A single amplifier • A preamplifier • IR emitters

2. The external consultants drew up the equipment list. What do you think the preamplifier is for?

____

____

Installation

1. On Figure 6-1 draw and label your recommended cabling infrastructure for the new audio system.

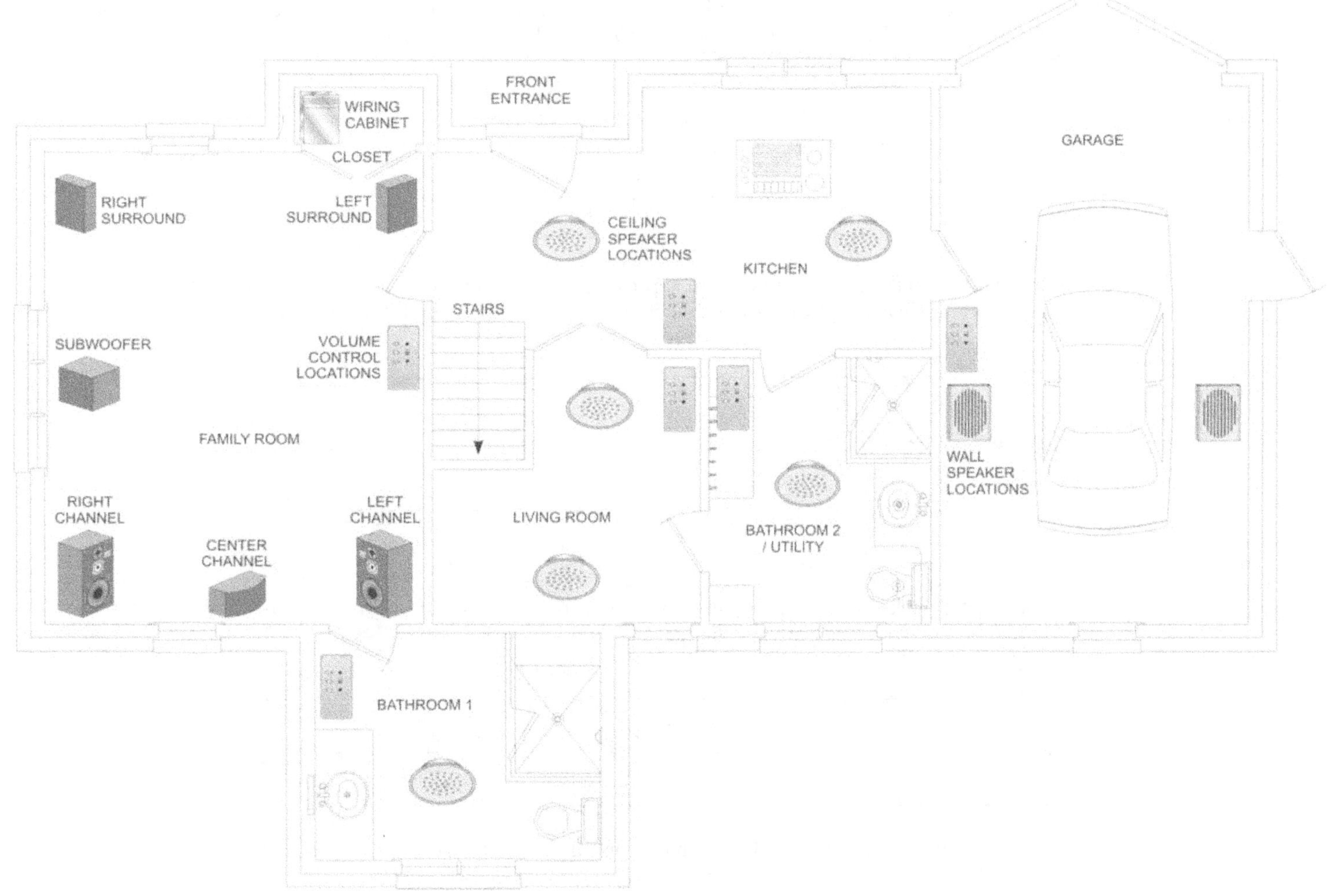

Figure 6-1: Your Recommended Cabling Infrastructure

2. SmartProperties has indicated that family members like to listen to very loud music. How will that influence the decision on the type of amplifier required for the new system?

____

____

3. What impact does the use of a single amplifier have on the proposed new system?

____

____

____

4. In-wall and in-ceiling speakers conserve floor space, offer good quality sound, and blend well into the room's decor. Therefore David and his interior designer have opted to install these types of speakers in all the rooms. Which of the following types of cables should SmartProperties use to comply with building safety standards?

____a. Special NECA/CL3-certified speaker wire
____b. Special NEC/CL3-certified speaker wire
____c. Special UL/CL3-certified speaker wire
____d. Special IEEE/CL3-certified speaker wire

5. The speakers in the main bedroom are approximately 120 feet from the amplifier installed in the utility room. What gauge of audio cable should you include in the bill of materials for the run?

____

Explain your answer:

____

____

6. What type of speaker system do you think SmartProperties should install in the home theater room to handle the low-frequency effects channel?

____

____

____

7. What type of speakers do you think SmartProperties should install in the bedrooms?

____

____

8. The interior designer working on the project team has selected the speakers for the various audio zones. The technical department at SmartProperties has the task of investigating three speaker ratings.

In Table 6-3, fill in the particular speaker rating with its corresponding description. Speaker ratings include sensitivity, impedance, and frequency response.

Table 6-3: Draft Design Plan

Speaker Rating	Description
	A measure of how well a speaker turns the power output of the amplifier into sound energy
	A measure of the opposition that a speaker has to ac and Dc current
	A measure of the audio frequencies that the speaker can produce

9. Indicate whether the statements about the installation of speakers in the home theater in Table 6-4 are true or false.

Table 6-4: Speaker Installation Statements (True/False)

True	False	Statement
		All speakers in a surround sound configuration should be spaced at different distances from the listener.
		The center speaker should be directly in front of the listener.
		Speakers that are installed behind the listener, should not be farther away than the front speakers.

Configuration

1. David wants to configure the system to drive a single speaker in the dining room with more power than the sum of the two original channels combined once installation and final trim out are completed. Which method should SmartProperties use to achieve this requirement?

____

2. David wishes to enable members of his family to remotely change the volume and control the audio sources. The cabling infrastructure has been installed. What are the next steps required to fulfill David's request?

____

____

____

____

3. Each piece of source equipment has its own IR-based remote control. What type of keypad should SmartProperties install to consolidate the multiple handheld remotes into a single control point in each of the audio zones?

____

____

____

Coax Cable Connections to the Integration Controller

OBJECTIVES

1. Identify and describe the function of coax cable connections between video sources.
2. Identify connection points between the video source equipment and the Integration Panel.
3. Interconnect the parts discussed within this lab procedure.
4. Establish a video connection between the DVD player and the LCD TV monitor.
5. Test the coax wall jack connections to the LCD TV monitor.

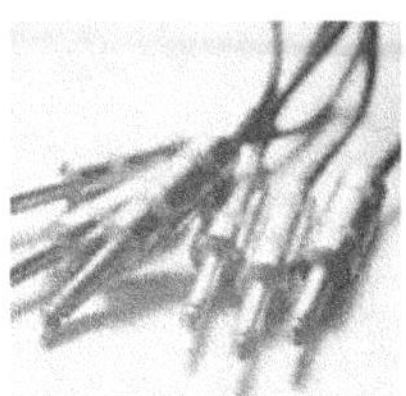

Audio/Video Fundamentals

RESOURCES

1. Marcraft Audio/Video Experiment Panel and Frame
2. Marcraft Integration Panel and Frame with some components mounted
3. Two RG-6 coax cables (length dependent upon location of the Integration Panel and the Audio/Video Panel)
4. Four twist-on F-connectors
5. RF modulator with power adapter
6. IR target
7. LED emitter
8. 3-foot male-male video cable
9. Coax cable coupler

TOOLS

1. RG-6 coax cable stripper
2. Philips screwdriver
3. Flat-tip screwdriver
4. Multimeter

DISCUSSION

Features offered by the whole-home video distribution system are as follows:

- The video distribution panel is the central hub for all video signals. It contains amplifiers, splitters, and modulators. It can distribute all home video sources to any television receiver in the home.

- RG-6 coaxial cable provides the input and output connections to the video distribution panel. Each room has dual coaxial cable wall outlets. The bottom connector on each outlet takes signals generated in that room from a VCR or video camera and feeds them back to an input on the distribution panel. This makes the signals available to all other TVs in the house.

- The outputs from the video distribution panel provide amplified modulated video inputs to each of the TV receivers in the separate rooms. The amplifiers are adjusted to provide the same video signal quality to each room.

- Modulators convert the video source signal (surveillance camera, satellite receiver, VCR, and DVD player) to an unused television channel for internal broadcasting on the video distribution system. This enables the home users to choose which video sources they wish to view by tuning the television receiver to the desired channel.

- A combiner consolidates video inputs from source equipment and connects them to the video input jack on the distribution panel.

PROCEDURES

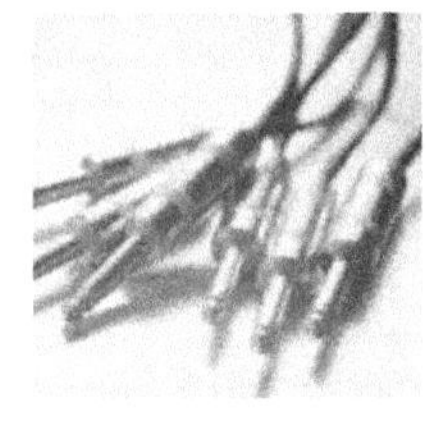

Audio/Video Fundamentals

Terminating a Coax Cable with Twist-on F-Connectors

1. Obtain two lengths of RG-6 coax cable.

NOTE: The length of these cables depends on the locations of the Audio/Video Panel and the Integration Panel. These cables will be used to connect the two panels.

2. Make a straight cut on the termination end of the coax cable.

3. Set the coax wire stripper to the **RG-6** setting (RG-59 and RG-62 are the same diameter.)

4. Adjust the stripper with the hex Allen wrench to the desired cable diameter and stripping requirements.

5. To obtain the best results, adjust the stripper to expose ¼ inch of the conductor and ¼ inch of the insulation.

6. Insert the cable into the stripper.

NOTE: The blades of the coax wire stripper will strip the end of the coax cable with a precise three-stage cut.

7. Strip the ends of the coax cable using a rotary motion (three to five full turns).

NOTE: Always turn the stripper the same direction. Do not cut all the way through the jacket as this will nick the shield. Flex the jacket for complete separation of the jacket.

8. Remove the coax cable from the stripper.

9. Inspect the quality of the work.

10. Check to ensure the center conductor and the insulation are not nicked or scored.

11. Check that stray strands from the braid are pushed away from the conductor.

12. Slide the twist **F-connector** onto the coax cable and tighten, as shown in Figure 7-1.

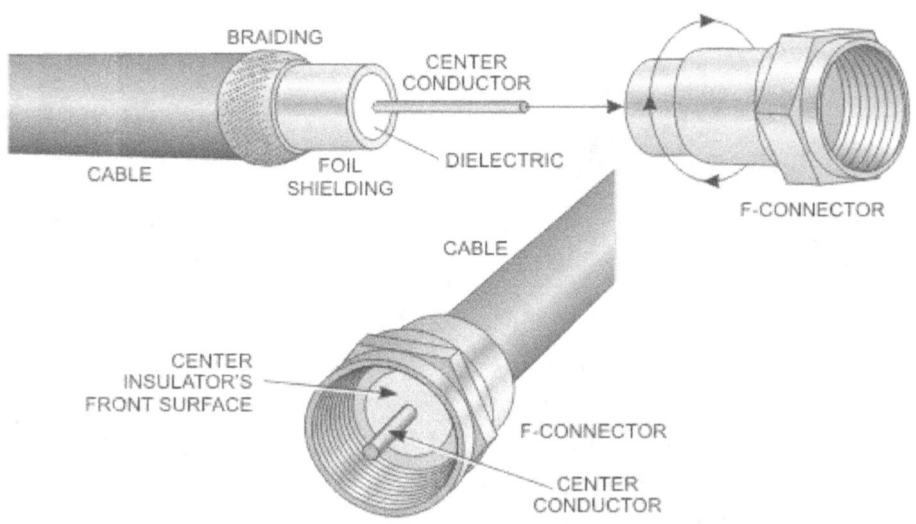

Figure 7-1:
Making a
Coax Cable

13. Verify that the center conductor is surrounded by insulation and that it is not touching the outer shell of the F-connector.

14. Similarly terminate the other end of the cable.

15. Now, terminate both ends of the second cable.

Checking the Coax Cable for Shorts

1. Perform a continuity test between the center conductor and the F-connector's outer shell with a multimeter.

2. Plug the red test lead for the multimeter in the **V/Ohm/mA** port of the multimeter.

3. Plug the black test lead for the multimeter in the **COM** port of the multimeter.

4. Set the multimeter dial to the **200** mark on the *Ohm* section of the multimeter.

5. Touch the meter leads together. See Figure 7-2.

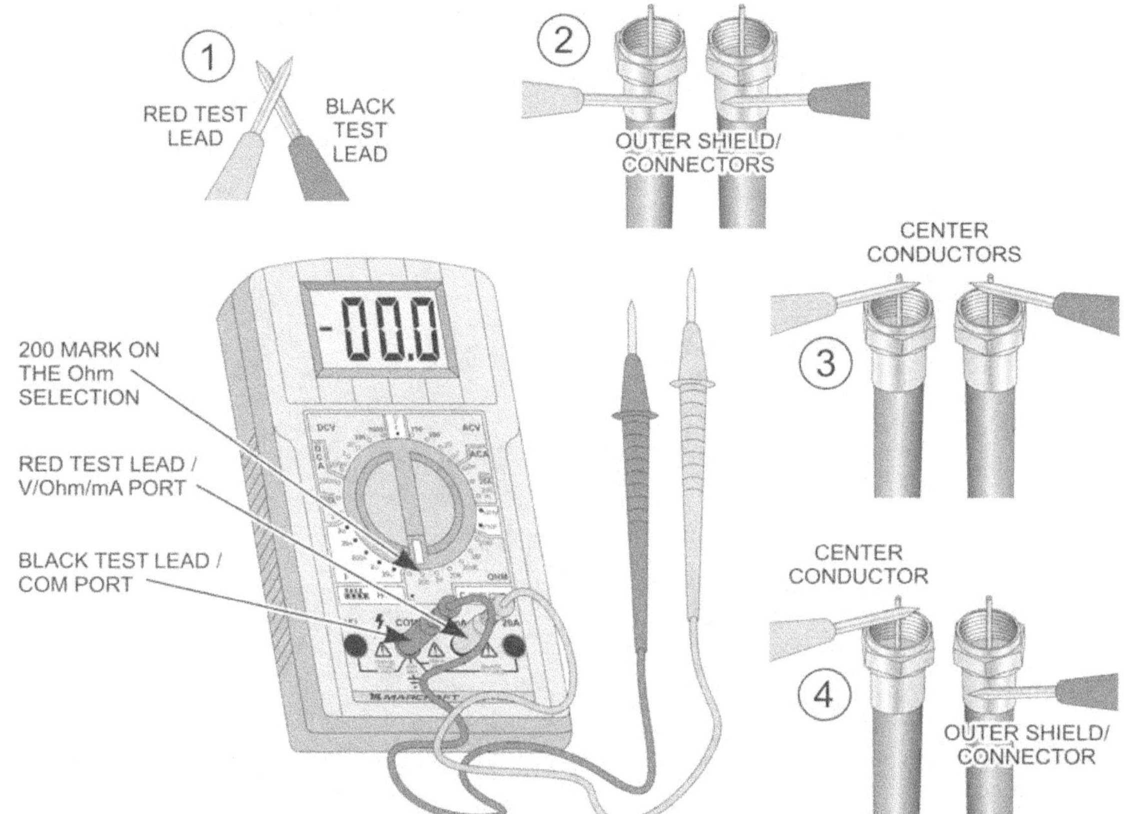

**Figure 7-2:
Coax Cable
Continuity Test**

6. Record the meter LCD reading in Table 7-1.

Table 7-1: Multimeter Readings for Coax Cable Continuity Test

Multimeter Test Leads		Test Reading
Meter Leads		
Outer Shell to Outer Shell		
Conductor to Center Conductor		
Outer Shell to Center Conductor		
Room A Left Coax Cable	Outer Shell to Outer Shell	
	Center Conductor to Center Conductor	
	Outer Shell to Center Conductor	
Room A Right Coax Cable	Outer Shell to Outer Shell	
	Center Conductor to Center Conductor	
	Outer Shell to Center Conductor	
Room B Left Coax Cable	Outer Shell to Outer Shell	
	Center Conductor to Center Conductor	
	Outer Shell to Center Conductor	
Room B Right Coax Cable	Outer Shell to Outer Shell	
	Center Conductor to Center Conductor	
	Outer Shell to Center Conductor	
Power Injector to IR Repeater Coax Cable	Outer Shell to Outer Shell	
	Conductor to Center Conductor	
	Outer Shell to Center Conductor	
IR Repeater to Splitter Coax Cable	Outer Shell to Outer Shell	
	Center Conductor to Center Conductor	
	Outer Shell to Center Conductor	
Student-Made Coax Cable 1	Outer Shell to Outer Shell	
	Center Conductor to Center Conductor	
	Outer Shell to Center Conductor	
Student-Made Coax Cable 2	Outer Shell to Outer Shell	
	Center Conductor to Center Conductor	
	Outer Shell to Center Conductor	

7. Touch one of the meter leads to the outer shell of the **F-connector** on the cable.

8. Touch the other meter lead to the outer shell of the other **F-connector**.

9. Record the meter reading in Table 7-1.

10. Touch one of the meter leads to the **center conductor** on one end of the cable.

11. Touch the other meter lead to the **center conductor** on the other end of the cable.

12. Record the meter reading in Table 7-1.

13. Touch one of the meter leads to the outer shell of the **F-connector** on the cable.

14. Touch the other meter lead to the **center conductor** on one end of the cable.

NOTE: Be careful not to touch the connector and the conductor at the same time. Touching the connector and conductor with the same meter lead will result in a short and give a false reading.

15. Record the meter reading in Table 7-1.

16. Perform the continuity test on the other coax cable, and record the resulting readings in Table 7-1.

Connecting the Integration Panel Wall Jacks to the Video Signal Splitter

1. Remove the faceplate from the wall outlet labeled **Room A**. See Figure 7-3.

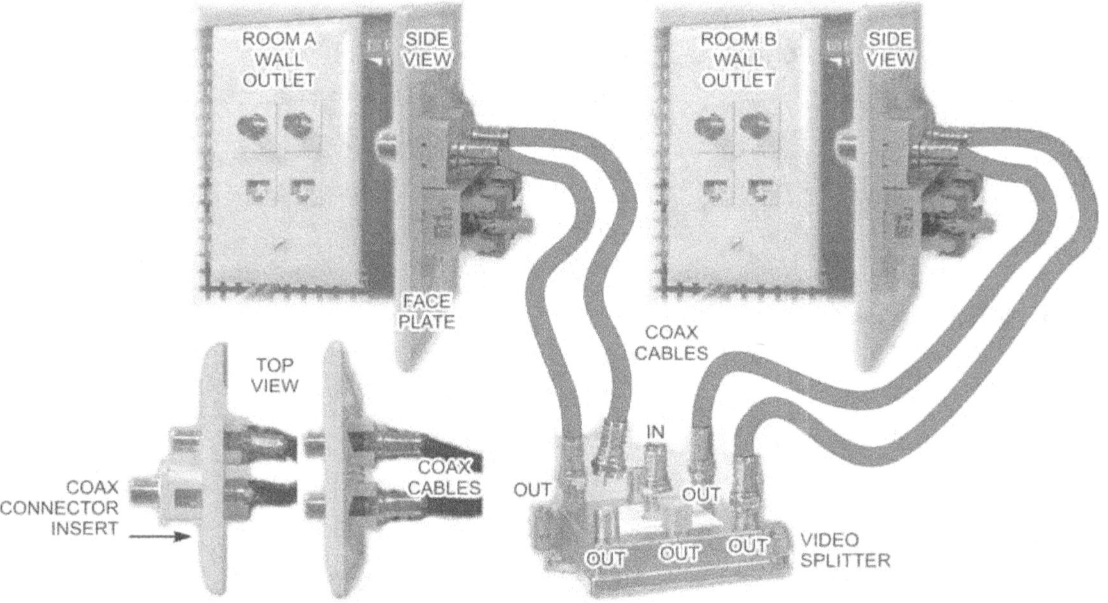

Figure 7-3: Connecting The Wall Outlet Coax Cable Jacks to the Splitter

NOTE: There are two coax cables in this electrical junction box. There may already be two CAT5 cables in the housing as well. Do not disturb the work of other students. The RG-6 cable is very stiff and inflexible. Take great care putting the wall outlet assembly back together. The connector inserts sometimes pop out of the faceplate. If the faceplate is damaged the inserts will not stay in the housing.

2. Perform a continuity test on both cables before assembly. Record the readings in Table 7-1.

3. Screw the coax cable F-connectors onto the mated end of the insert on the wall outlet faceplate.

4. Replace the faceplate assembly on the wall outlet housing, being careful to coil all of the cables into the housing.

5. Remove the faceplate from the wall outlet labeled **Room B**.

6. Perform a continuity test on both cables before assembly. Record the readings in Table 7-2.

7. Screw the coax cable F-connectors onto the mated end of the insert on the wall outlet faceplate.

8. Carefully replace the faceplate for **Room B** on the wall outlet housing.

9. Connect the coax cables in the **Room B** wall outlet.

10. Connect the four coax cables from the Room A and Room B wall outlets to the **six-way splitter** in the Integration Controller housing. Use any of the six connectors labeled **OUT**. Refer to Figure 7-3 for the reference picture of the video splitter.

Connecting the Video Splitter to the Power Injector Assembly

1. Remove the faceplate for the wall outlet labeled **Power Injector**. See Figure 7-4.

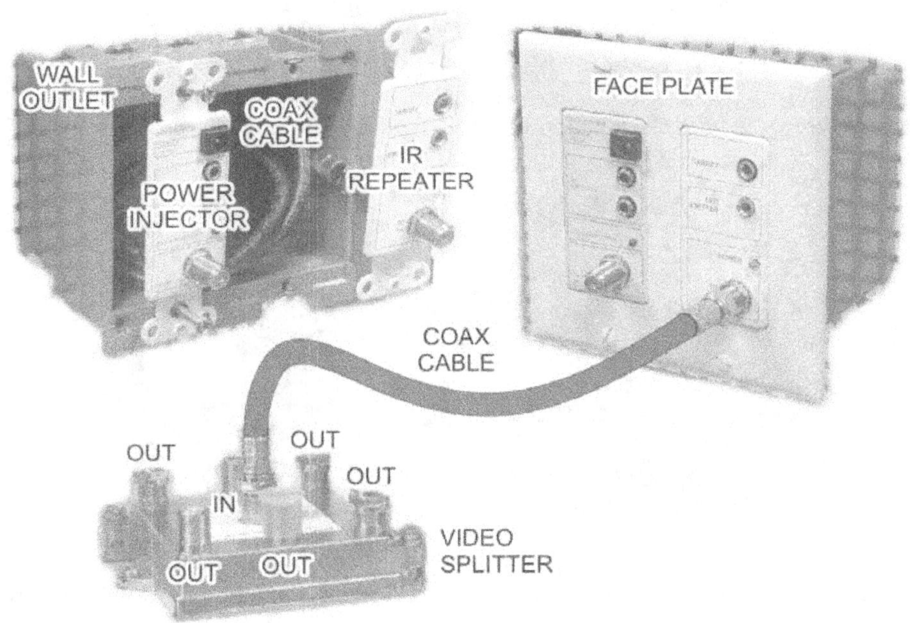

**Figure 7-4:
Connecting the
Power Injector
and IR Repeater
Assemblies to the
Video Splitter**

2. Remove the power injector and the IR repeater from the wall outlet housing.

3. Perform a continuity test on the coax cable used to connect the power injector to the IR repeater. Record the readings in Table 7-1.

4. Perform a continuity test on the coax cable used to connect the IR repeater to the video splitter. Record the readings in Table 7-1.

5. Connect a short coax cable between the power injector and the IR repeater.

6. Carefully replace the power injector and IR repeater assemblies into the wall outlet housing.

NOTE: The coax cable may need to be formed into the correct shape while placing the assemblies into the housing. Use great care not to damage the power injector or IR repeater.

7. Put the faceplate onto the wall outlet housing.

8. Connect the cable coming from the video splitter to the port labeled **TO TV** on the IR repeater.

9. Connect the other end of the coax cable to the port labeled **IN** on the video splitter.

Setting the Dip Switches Inside the RF Modulator

1. Review the documentation provided with the RF modulator.

NOTE: Do not touch anything inside the RF modulator housing except the dip switches. This product contains an electronic circuit board and is susceptible to static charge damage.

2. Make sure the power adapter is removed from the RF modulator.

3. Remove the two screws and the black plate from the RF modulator housing. See Figure 7-5.

**Figure 7-5:
Setting the Dip
Switches on the RF
Modulator**

4. Set dip switches 1 and 2 to the **on** position (down or in the direction of the arrow labeled "ON" on the dip switch module).

NOTE: The sum of the numbers 80 and 40 from the dip switches equals 120. This is the channel on which the TV will receive the modulated signal.

5. Ensure that the remaining dip switches are in the **off** position (up or in the opposite direction of the arrow labeled "ON").

6. Put the back plate on the RF modulator and secure it with the screws.

Connecting the RF Modulator to the Power Injector

1. Refer to Figure 7-6 while connecting the video system devices to the RF modulator and power injector.

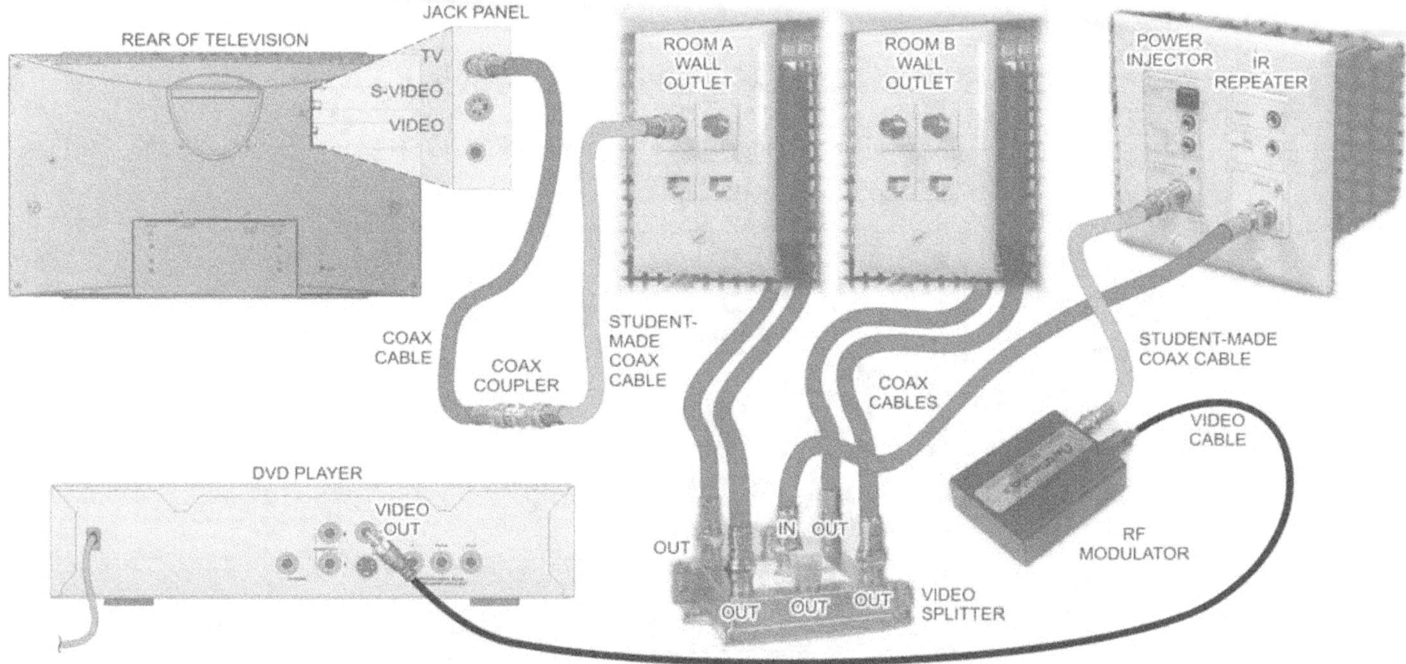

Figure 7-6: Connecting the RF Modulator and DVD Player to the Integration Panel

2. Disconnect the video out signal cable for the TV from the DVD player.

3. Connect a spare video signal cable from the DVD player to the RF modulator.

4. Connect one of the student-made cables to the RF modulator.

5. Connect the other end of the coax cable to the power injector.

6. Plug the 12 Vdc power adapter into the RF modulator.

7. Plug the 12 Vdc power adapter into an electrical outlet.

8. Connect the second student-made coax cable to the coax coupler attached to the cable running to the TV input on the back of the LCD flat-screen TV.

9. Connect the other end of the coax cable to the left coax jack of the Room A wall outlet.

Tuning the LCD TV Monitor

1. Review the section of the LCD TV owner's manual pertaining to Automatic programing.

2. Turn on the LCD TV monitor.

3. Turn on the DVD player.

4. Insert the **Avia II Guide to Home Theater** DVD into the DVD player.

5. With the LCD TV monitor remote, push **Enter** to select **ENGLISH** as the default language. See Figure 7-7.

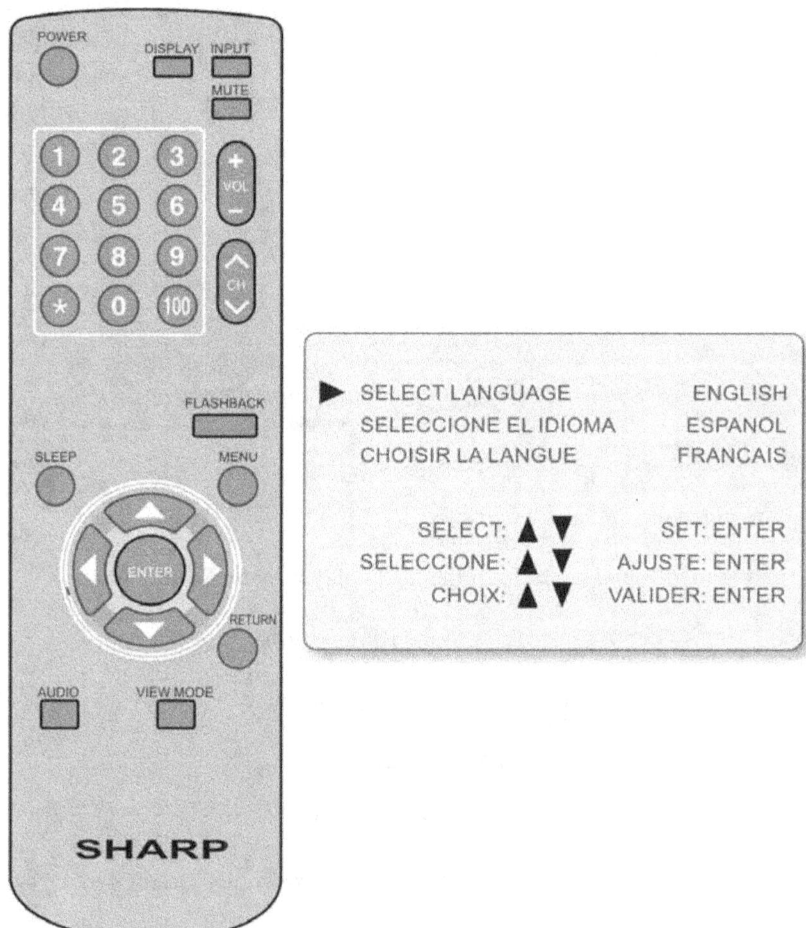

**Figure 7-7:
Setting the Tuner
Mode to Cable**

6. Using the control buttons surrounding the Enter button, choose **AUTO PRESET CH (CATV)** and press **Enter**. See Figure 7-8.

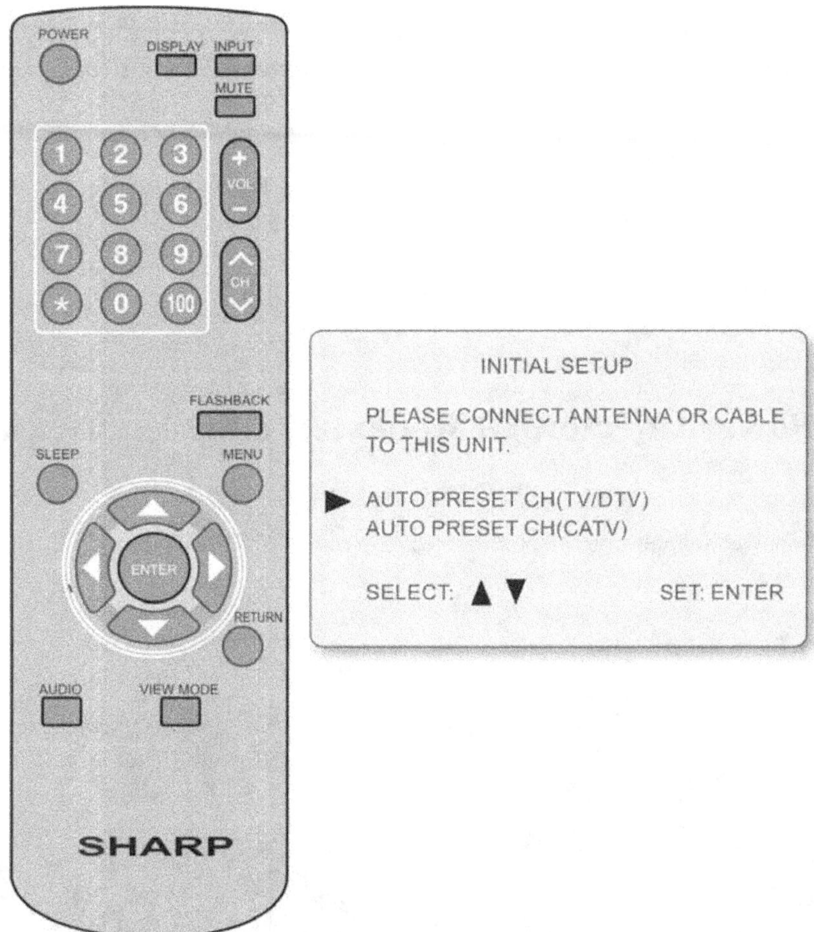

**Figure 7-8:
Performing
Auto Program**

7. The television will now begin cycling through each channel as it searches for a video input signal.

8. Once the PRESET scan is finished, the TV will display the channels found receiving a signal.

NOTE: The end result of the scan should be one (1) channel being displayed, channel 120; the only channel receiving a video input signal. No other channels should have been PRESET during this process.

Checking the Remaining Three Coax Wall Jacks

1. Remove the coax cable from the left coax wall jack connector on the Room A wall outlet.

2. Connect the coax cable to the right coax wall jack connector on the Room A wall outlet.

3. Check to see that the LCD TV monitor still has picture.

4. Repeat this check for the two remaining coax wall jacks on the Room B wall outlet.

NOTE: If the picture on the LCD TV monitor is not present during the coax wall jack testing, inspect all the cables involved. If no connections can be found visually, re-entry into one of the wall outlets may be necessary.

5. Turn off the LCD TV monitor.

6. Remove the Avia Guide to Home Theater DVD from the DVD player.

7. Turn off the DVD player.

Testing the Power Injector, IR Repeater, LED Emitter, and Target

1. Plug the 15 Vdc power adapter into the power injector. See Figure 7-9.

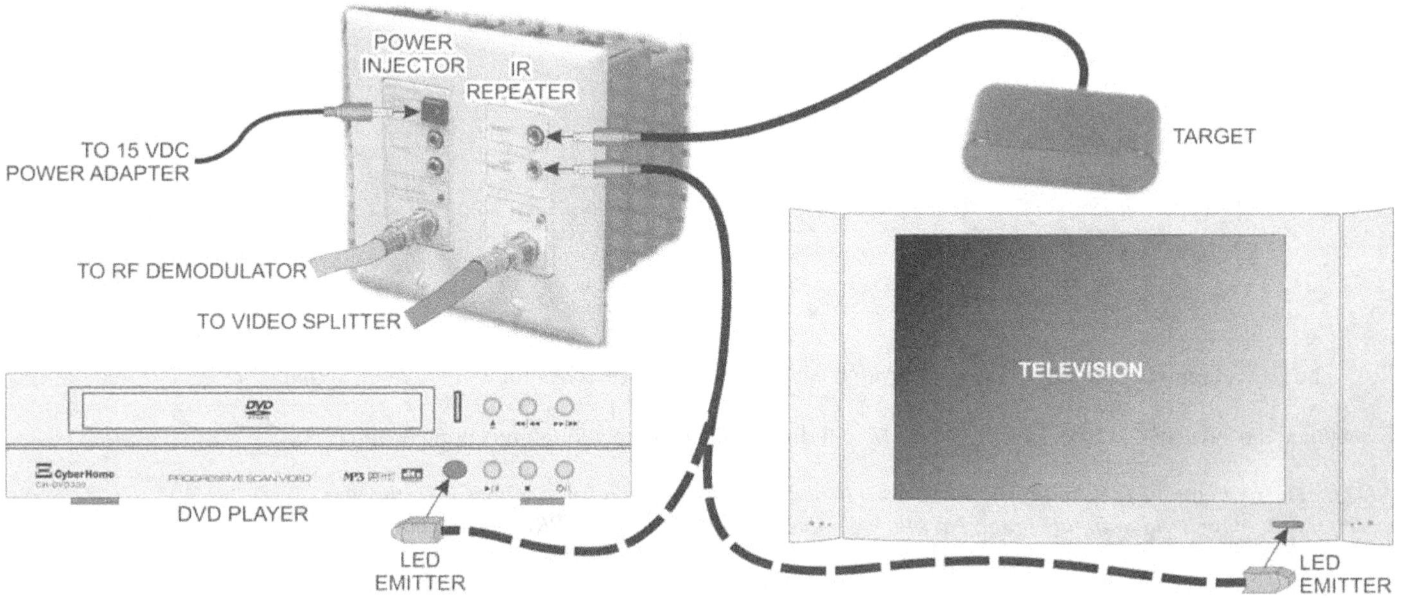

Figure 7-9: Testing the LED Emitter and Target

2. Plug the 15 Vdc power adapter into an electrical outlet.

3. Plug the target device into the **TARGET** port on the IR Repeater.

4. Place the target device on top of the double-gang electrical junction box.

5. Plug the LED emitter into the **LED EMITTER** port on the IR repeater.

6. Route the LED emitter to the front of the audio/video panel.

7. Do not remove the protective cover paper from the adhesive on the LED emitter.

8. Verify that the power LEDs are on for the power injector, IR repeater and target device.

9. Hold the LED emitter in front of the LCD TV monitor.

10. Aim the remote for the LCD TV monitor at the target device.

NOTE: Hold the remote control away from the LCD TV monitor, behind the monitor if possible, to ensure the LCD TV monitor responds to the LED emitter and not to the remote control.

11. On the remote control for the LCD TV monitor, press the **Power** button.

12. Hold the LED emitter in front of the DVD player.

13. Aim the remote control for the DVD player at the target device.

NOTE: Hold the remote control away from the DVD player, behind the DVD player if possible, to ensure the DVD player is responding to the LED Emitter and not the remote control.

14. On the remote control for the DVD player, press the **Power** button.

15. On the remote control for the DVD player, press the **Open/Close** button to open the DVD player tray.

16. On the remote control for the DVD player, press the **Open/Close** button to close the DVD player tray.

17. Use the target device and both remote controls to turn off the power for the DVD player and the LCD TV monitor.

Speaker Connections to the Integration Controller

OBJECTIVES

1. Identify and describe the function of stereo cable connections between audio sources.
2. Identify connection points between the audio source equipment and the Integration Panel.
3. Interconnect the parts discussed in this lab procedure.
4. Establish an audio connection between the DVD player, the dual-source, four-zone audio distribution system, and the wall speakers.
5. Test the four separate zones for sound connections to the wall speakers.

Audio/Video Fundamentals

RESOURCES

1. Marcraft Audio/Video Experiment Panel and Frame
2. Marcraft Integration Panel and Frame with some components mounted
3. Eight stereo cables (length dependent upon locations of the Integration Panel and the Audio/Video Panel)
4. Sixteen wire nuts

TOOLS

1. Cable stripper
2. Philips screwdriver
3. Flat-tip screwdriver

DISCUSSION

Audio Distribution System

The audio distribution system shares some design features with the video distribution system. Audio distribution requires audio source equipment located in a central location to be shared by different users in various locations called zones. A zone consists of a group of speakers in one area of the home. Audio sources can be designed with either a single-zone or multizone distribution plan. Single-zone design enables all areas to hear the same audio source at any interval of time. Multizone design permits different users in different areas of the home to listen to different sources independently.

Volume controls and source equipment keypads are usually mounted in wall outlets. They're wired into the system so they can control the sound in a room or the entire home. Users can employ keypads to select the source and control the volume either manually or with handheld infrared remote units.

Audio distribution systems use various wire and cable types for connecting components including Category 5 UTP cable, speaker cable, and shielded audio patch cables. Source audio equipment is usually located in a "headend" location where all the audio and video equipment is installed in a shared audio/video equipment rack.

Selection of audio source equipment and controls depends upon the size of the home, the number of rooms, and the number of zones.

Audio/Video Fundamentals

PROCEDURES

1. Refer to Figure 8-1 while connecting the wall speakers to the Integration Panel.

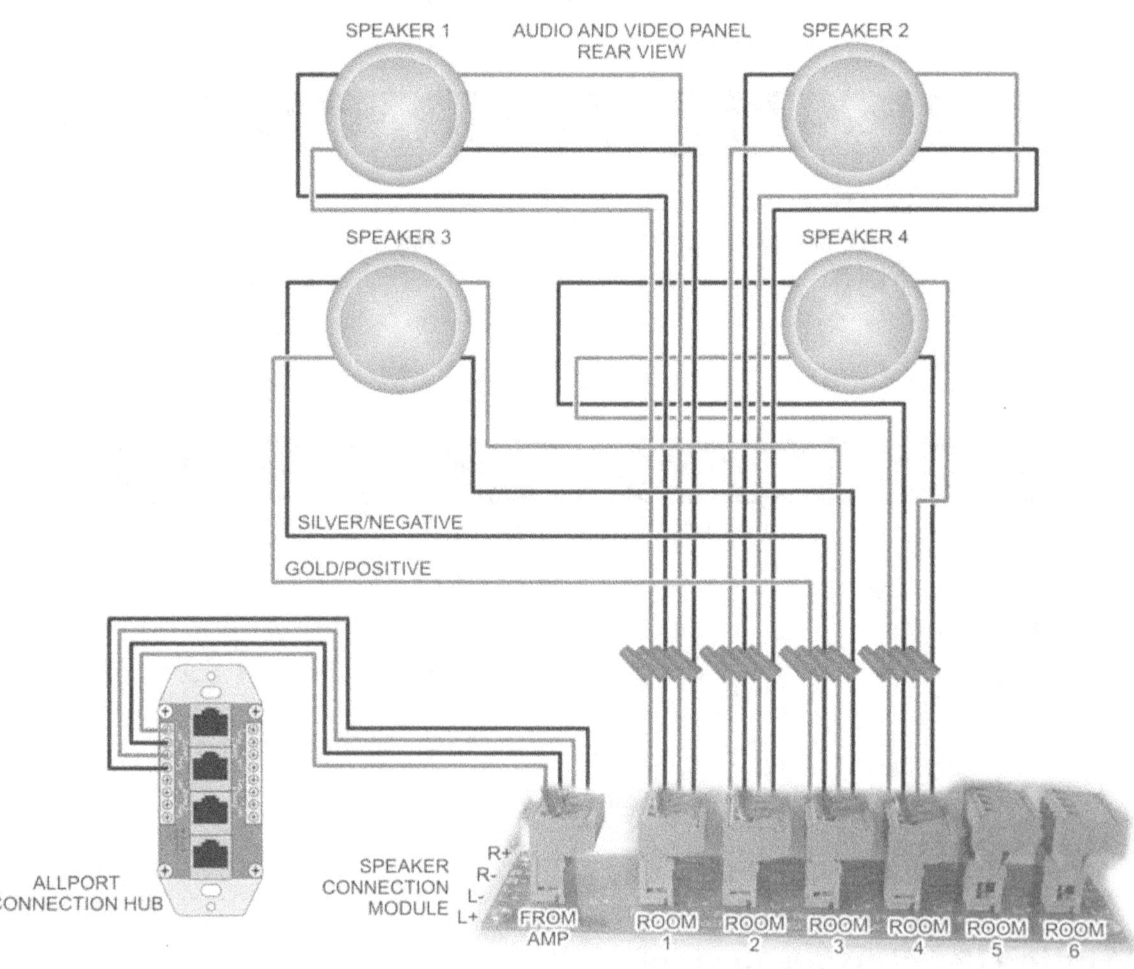

Figure 8-1: Speaker Connections to the Integration Panel

2. Obtain ten lengths of stereo cable.

NOTE: The lengths of the stereo cable are dependent upon the locations of the Integration Panel and the Audio/Video Panel.

3. Note the polarity markings on the speaker connection module.

4. Try a dry fitting of one of the stereo modular connectors to the speaker connection module to ascertain proper orientation.

5. Note how the stereo wires must be installed to the modular connector to maintain proper polarity.

6. Remove the left + (gold) and - (silver) stereo cables for Zone 1 from the dual-source, four-zone audio distribution system.

7. Use wire nuts to connect one of the lengths of stereo cable to the cables removed from the dual-source, four-zone audio distribution system. Maintain conductor color (gold to gold and silver to silver).

8. Route the cable back and into the top third hole of the controller housing.

9. Checking for proper polarity, connect the cable to a stereo modular connector's L+ and L- terminals.

10. Remove the right + (gold) and - (silver) stereo cables for Zone 1 from the dual-source, four-zone audio distribution system.

11. Using wire nuts to connect a second length of stereo cable to the cables removed from the R+ and R- channels of Zone 1 from the dual-source, four-zone audio distribution system. Maintain conductor color (gold to gold and silver to silver).

12. Route the cable back and into the top third hole of the controller housing.

13. Checking for proper polarity, connect the cable to a stereo modular connector's R+ and R- terminals.

14. Connect this stereo modular connector to the terminal labeled **ROOM 1** on the speaker connection module.

15. Repeat steps 1–14 for the speakers of Zone 2, Zone 3, and Zone 4, using the modular connector terminals labeled Room 2, Room 3, and Room 4 respectively.

Audio Connections to the Integration Controller

16. Refer to Figure 8-1 while connecting the audio system to the Integration Panel.

17. Connect a length of stereo cable to the dual-source, four-zone audio distribution system **L+** and **L-** Zone 1 connector.

18. Route this cable back and into the top third hole of the controller housing.

19. Checking for proper polarity, connect the cable to a stereo modular connector's **L+** and **L-** terminals.

20. Connect a length of stereo cable to the dual-source, four-zone audio distribution system **R+** and **R-** Zone 1 connector.

21. Route this cable back and into the top third hole of the controller housing.

22. Checking for proper polarity, connect the cable to a stereo modular connector's R+ and R- terminals.

23. Connect this stereo modular connector to the terminal labeled **FROM AMP** on the speaker connection module.

24. Turn on the LCD TV monitor.

25. Turn on the DVD player.

26. Put the Avia Guide to Home Theater DVD into the DVD player.

27. Power up the dual-source, four-zone audio distribution system.

28. On the Zone 1 keypad of the Audio/Video Panel press the **ON/OFF** button and turn the keypad **on**. See Figure 8-2.

29. Press the **B** button to select the channel for the DVD player.

30. Ensure that the DVD player is in the play mode and sending an audio signal.

31. Adjust the volume on the keypad up or down as necessary.

32. Turn the *Zone 1* keypad **off**.

33. Repeat steps 25 through 32 for the **Zone 2**, **Zone 3**, and **Zone 4** keypads.

34. Turn off the LCD TV monitor.

35. Remove the Avia Guide to Home Theater DVD from the DVD player.

36. Turn off the DVD player.

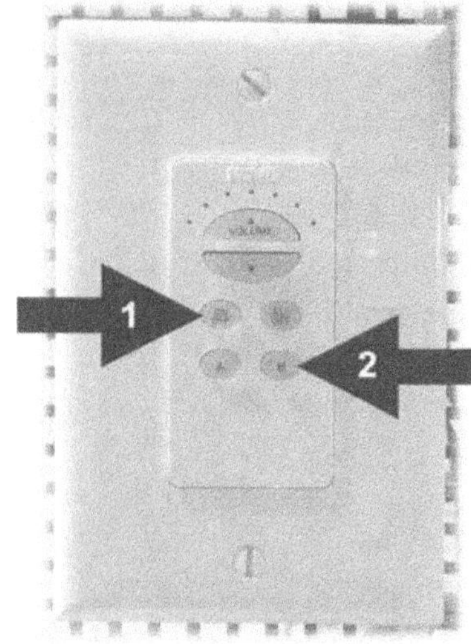

Figure 8-2: Dual-Source, Four-Zone Audio Distribution System Keypad

NOTE: If at any time during the speaker testing process, a speaker does not produce sound, make sure the DVD player is in the play mode. If the DVD player is playing and no sound is heard, check all stereo cables involved for the speaker in question.

LAB QUESTIONS

Feedback

1. What capability does the video distribution panel provide?

2. What type of connector is used for the input and output connections to the video distribution panel?

3. What is the function of a modulator?

4. How is the audio distribution system similar to the video distribution system?

5. Where are volume controls and source equipment keypads usually mounted?

6. What wire and cable types are used for connecting components in the audio distribution system?

7. What is a factor in the selection of audio source equipment and controls?

Audio and Video Objective Map

There are two levels of expertise proposed for those workers who install electronics cables in residences and interconnect electronics communications, computer, control or entertainment equipment. **RESI, the Residential Electronics Systems Integrator**; and the **Master RESI, Residential Electronics Systems Integrator.**

The RESI integrator is proficient in pre-wiring for home theater and telecommunications equipment interconnection. He/she will install wiring for antenna, satellite, cable TV, wireless broadband, audio/video entertainment and computer equipment/networking. He/she will have skills required for low voltage wiring. Workers who hold the RESI certification will likely also hold one or more of the 5 specialty endorsements depending on the area of expertise that worker requires.

The MASTER RESI will be proficient at all of the RESI skills and knowledge as well as in planning and designing electronics and communications equipment systems and layout for new and existing construction. The RESI worker is capable of designing the entire system and network for audio, video, data and control of security and environment. He/she also is capable of troubleshooting and debugging the system and planning installation or modifications. The MASTER RESI has extensive knowledge of the operation and technology and is proficient in each of the basic five subcategories of residential electronics.

Domain 1.0   Signals

Content may include the following:

1.1 Describe telephone system signal types

1.2 List audio signal sources and types

1.3 Compare video signals with radio frequencies and voice

1.4 Differentiate between video, audio and data Signals

Domain 2.0   Amplifiers

Content may include the following:

2.1 Compare power amplifiers used in residential systems with other types of audio amplifier circuits

2.2   Interpret an amplifier's specifications sheet

2.3 Explain the Dolby sound system, Dolby Digital Ex. and DTC and list the advantageous features of each

2.4   Explain the features and operation of Pro Logic and compare with basic audio amplifier sound processing

2.5 Describe CDs and DVD inclusion in the audio/video system

Domain 3.0   Speakers

Content may include the following:

3.1 Compare various freestanding speakers commonly utilized in home theater systems

3.2 Differentiate between ordinary 2 channel audio and Surround systems

3.3 Explain the purpose of subwoofers, their frequency range and styles

3.4 Explain the advantages of bookshelf speaker units

3.5 Describe the pre-wiring of In-Wall speakers, mounting and connections

3.6 Define impedance and explain its importance in matching

3.7 Explain speaker stiffness and its effect on audio balance

3.8 Explain speaker mass and its effect on audio frequencies

3.9 Explain damping factor in speakers

Domain 4.0   Speaker Cabling

Content may include the following

4.1 Explain ways that wrong wire gauge can adversely affect the sound system

4.2 Describe connector requirements for multi-room In-Wall speaker

4.3. Compare speaker terminal connections, spring clips, binding posts, etc.

4.4 Explain the value of oxygen-free copper wire for audio systems

4.5 Define: Transient Distortion, Wow and Flutter

4.6 Describe advantages, compare costs and quality with wired speakers

Domain 5.0   Audio Systems Basics

Content may include the following:

Domain 6.0   Surround Sound

Content may include the following:

6.1 Explain Surround Sound basics

6.2 Sketch the placement of surround sound speaker units

6.3 Describe how to rewire to allow room areas use of surround equipment

6.4 Explain DTS and its advantages

Domain 7.0   Acoustics

Content may include the following:

7.1 Acoustics – define and present examples of good and bad

7.2 Explain acoustic resistance and resonance

7.2 Explain causes of detrimental sound reflection

7.3 Explain sound refraction and discuss causes

7.4 Explain sound diffraction and explain detrimental aspects

7.5 Describe low frequency effects an area may exhibit

Domain 8.0   Video

Content may include the following:

8.11 Explain importance of seeking customer choices for products and in-home positioning

8.12 Describe display maximization – color temperature/balance

8.13 Describe various types of wireless control of audio and video equipment using remote hand units, LED and RF sending/receiving devices

Domain 9.0   Display Devices

Content may include the following:

9.1 Describe the operation and handling precautions for CRT displays

9.2 Describe CRT, LED, LCD and other projection TV technologies

9.3 Compare Monitors with TV receivers and list advantages

9.4 Compare DLP/LCD/LcoS/Plasma technologies and list advantages of each

9.5 Describe projectors/screens used in home theater applications

9.6 Describe actuators and remote control for motorized screens

9.7 Discuss scanning – interlacing, progressive and the features of both

9.8 Explain the purpose and technology of De-Interlacing and Line Doubling

Domain 10.0   Home Theater Systems

Content may include the following:

10.1 Describe audio channel selection on a receiver/amplifier

10.2 List advantages of remote wireless control of whole-house electronics from the home theater primary viewing area

10.3 Summarize modern displays and speaker aesthetics for residences

10.4 Discuss home theater seating concepts

10.5 Explain requirements of cabling, speakers and display units when retrofitting a home

10.6 Sketch the components for a motorized projection screen

Domain 11.0   Off-air Antennas

Content may include the following:

11.1 Explain off-air installation basics

11.2 Identify types of antenna-dish components and mounts

11.3 Discuss logical methods of troubleshooting dish and antenna problems

11.4 Compare analog/digital/HDTV Broadcast Signals

11.5 Describe the required small dish system installation and programming procedures that must be followed and the configuration required prior to customer subscription access.

Domain 12.0   Cable TV

Content may include the following:

12.1 Describe cable TV connections, ground blocks and wiring from street to home
and entry interconnection

12.2 List minimum signal levels common for CATV systems and the use of line amplifiers where weak signals exist

12.3 Define DSL, B-VoIP, PPV and Telephone Services. List major advantages of
each

Domain 13.0   Distribution Systems

Content may include the following:

13.1 Illustrate home run and daisy chain wired signal distribution systems

13.2 Illustrate and compare wireless distribution systems with wired systems

13.3 List common usages for RG 59, RG 62, RG6, CAT5e/6 and fiber cabling and the
advantages of each

13.4 Diagram residential signal distributions equipment interconnection

13.5 Describe how a home computer network can be used to remotely control home electronics

13.6 Explain how individual room & areas can be control from multiple locations

13.7 Describe pre-wiring and retro wiring methods, wall fish, attics, crawl spaces, etc.

13.8 Describe 70 volt sound distribution technology and indicate applications where it is desirable

13.9 List Internet resources that may be included in home networks

13.10 Explain the use of splitters, diplexers, taps, fittings and outlets

13.11 Explain how plastic fiber optics can be utilized in the home systems

Domain 14.0   Troubleshooting

Content may include the following:

14.1 Describe common technical problems in home theater systems

14.2 Describe methods and equipment used to troubleshoot signal systems or to substitute or detect systems signals

14.3 Explain the usage of signal generators/ TDR and DMM

14.4 List signal problems from external sources or those caused by the A/V system components

14.5 List tools and test equipment used for installation work in homes

14.6 Describe methods and equipment used to maximize A/V Equipment capabilities

14.7 Describe procedures for accessing and resolving in-wall equipment/cabling problems

14.8 Present an example of customer equipment faults leading to dissatisfaction of the electronics work and potential loss of income for the installing dealer

14.9 Explain the grounding process and its importance in reducing ground loops as well as customer lighting/surge concerns

Glossary

10Base-T: An Ethernet specification defined by the IEEE 802.3 committee officially as specification 802.3i. The specification describes a 10 Mbps wire speed using baseband information. The specification requires a star topology using Category 3, 4, or 5 unshielded twisted pair (UTP) cable.

100Base-T: An Ethernet IEEE 802.3u specification with a wire speed of 100 Mbps using Category 5 or 5e UTP cable.

1000Base-T: An IEEE 802.3ab standard for Gigabit Ethernet. Gigabit Ethernet runs at 1,000 megabits per second using Category 5e cable.

A

AC cable: Armored cable is identified by the code name AC cable. AC cable is often referred to as *BX*, although this is the trademark of a specific manufacturer. Armor clad and metal clad (MC) might sound like the same cable, but some important differences exist. AC cable uses the interior bond wire, in combination with the exterior interlocked metal armor, as the equipment grounding means of the cable. MC cable, on the other hand, is manufactured with a green insulated grounding conductor, and this conductor, in combination with the metallic armor, comprises the equipment ground. AC cable can have up to four insulated conductors only; a fifth insulated conductor is allowed by UL, if it is a grounding conductor. Each conductor in AC cable is paper-wrapped.

AC-3 (Audio Coding-3): Dolby's digital audio data compression algorithm adopted for HDTV transmission, laserdiscs, and CDs for 5.1 surround sound multichannel home theater use.

ADSL terminal unit (ATU): The engineering term used for describing an ADSL modem. The ADSL terminal unit is also known as the ADSL transceiver unit, ADSL transmission unit, and ADSL termination unit. The ATU-R (remote) unit is installed in the home, and the ATU-C (central) office unit is installed in the local telephone company central office.

Advanced Television Systems Committee (ATSC): The group that developed voluntary national standards for high-definition television (HDTV), standard-definition television (SDTV), data broadcasting, multichannel surround sound audio, and satellite direct-to-home broadcasting.

air handler: The subassembly contained in an HVAC air distribution system that typically houses items such as the supply fan(s), return fan(s), heating coil, cooling coil, and filters. Its purpose is to provide the pressure necessary to get the airflow through the entire system, and at the desired temperature and humidity.

alarm system: A dedicated electronic system using a central controller and distributed sensors that detects unauthorized intrusion into a home interior, or protected external areas. It is typically used with door and window sensors, glass break sensors, motion detectors, and occasionally closed circuit television. When armed by the user, the system detects intrusion from the outside and can react with responses that vary from the activation of interior sirens, bells, and lighting, to the notification of a central alarm monitoring system.

American National Standards Institute (ANSI): A private, nonprofit, nongovernmental national organization that serves as the primary coordinator of standards within the United States, in relation to ISO standards.

American Wire Gauge (AWG): Standard measuring gauge for nonferrous conductors. Gauge is a measure of the diameter of the conductor. The higher the AWG number, the thinner the wire.

ampere: A measuring unit for the rate of electron flow or current in an electrical conductor. One ampere of current represents one coulomb of electrical charge (6.24×10^{18} charge carriers) moving past a specific point in one second. The ampere is named after Andre Marie Ampere, a French physicist (1775-1836).

amplifier bridging: A technique used to configure a two-channel stereo amplifier to drive a single load (speaker) with more power than the sum of the two original channels. An amplifier running in bridged mode has a single output channel to which a load (speaker) can be connected.

amplitude: The magnitude of a signal in voltage or current. It's frequently expressed in terms of peak, peak-to-peak, or RMS.

analog: A term used in telecommunications technology to describe a continuously variable waveform that is analogous to the electrical signal produced by the human voice when processed by a microphone and audio amplifier. Also, it's a signal in which a base carrier's alternating current frequency is modified in some way, such as by amplifying the strength of the signal, or varying the frequency to add information to the signal (also called *amplitude modulation*). Broadcast and telephone transmissions have conventionally used analog technology. An analog signal can be represented as a series of sine waves. The term originated because the modulation of the carrier wave is analogous to the fluctuations of the human voice or other sound that is being transmitted.

anti-siphon valve: A control valve with a built-in atmospheric vacuum breaker (backflow preventer). Most commonly used in residential irrigation systems.

apparent power: In an AC circuit apparent power is the assumed value of P = E (volts) and I (amperes) and is expressed as volt-amperes (VA). Apparent power is the theoretical power consumed in a circuit, while neglecting the power factor. (See "true power.")

asymmetric digital subscriber line (ADSL): Asymmetric DSL is a class of DSL service described in ANSI TI.413-1998 that offers higher download speeds with a slower upstream speed. The actual network bandwidth a customer receives from a DSL service in the home depends on the total length of the telephone local loop distance to the local office.

Atmospheric Vacuum Breaker (AVB): A type of backflow preventer. (See "valves.")

Audio Engineering Society (AES): Founded in 1949, the largest professional organization for electronic engineers and all others actively involved in audio engineering. It's primarily concerned with education and standardization.

audio/video (A/V) receiver: A component of a home audio/video system. A/V receivers have an integrated AM/FM tuner, and process from 4 to 6 channels with amplifiers capable of driving multiple speaker channels. They have built-in surround sound decoders that separate the sound signal from a video source (like a DVD player) into several channels, then route it to several speakers.

B

backflow prevention valve: A valve used with irrigation systems that prevents potable water from being contaminated with nonpotable water in the event a cross-connection exists during a condition of backflow.

backwash valve: A valve that reverses the flow of water in a swimming pool filter during the filter cleaning cycle. Modern backwash valves are molded of temperature- and chemical-resistant chlorinated polyvinyl chloride (CPVC) material.

bandwidth: The transmission capacity of a medium expressed as a range of frequencies in hertz (Hz). A greater bandwidth indicates the capability to transmit a greater amount of data over a given period of time.

Bayonet Neil Concelman (BNC) connector: A coaxial cable connector that, in its male form, has a center pin (bayonet) connected to the center conductor and a metal tube connected to the cable shield. A rotating ring surrounds the tube and contains small holes that mate with projections on the female connector to make the connection. The connector was developed in the late 1940s and is named after the creators — Amphenol engineer Carl Concelman and Bell Labs engineer Paul Neill. Neill designed the N-type connector, and Concelman designed the C-type connector. The BNC is a hybrid N/C-type with a mechanical extra appropriately called a *bayonet*. BNC connectors are often identified using other terms of unknown origin, such as British naval connectors, bayonet navy connectors, or bayonet nut connectors.

bimetal element: A type of temperature sensor used in older thermostats containing two strips of different metals that are attached together. As the room temperature changes, the lengths of the metal strips change. Because the metals are different and change their lengths at different rates, they are used to operate thermostat switches that react to temperature changes.

bit (binary digit): The smallest component of information in a binary notation system. A bit is a single 1 or 0.

bit rate: The number of binary digits transmitted over a medium per unit of time. Bit rate is specified as the number of bits transmitted in one second (bps).

Bluetooth: A wireless technology standard designed to connect one device to another with a short-range radio link. This standard has evolved from early work and engineering studies performed by Ericsson Mobile Communications in 1994. The IEEE Project 802.15.1 has derived a Wireless Personal Area Network (WPAN) standard based on the Bluetooth v1.1 Foundation specifications.

booster pump: A device that increases the water pressure in a system where some pressure already exists.

branch circuits: The circuits in a house that branch from the service panel to boxes and devices.

breaker: A switch-like device that opens a circuit if the current to the branch circuit exceeds the specified rating of the device.

Broadband over Power Line (BPL): A system that proposes to use the electric power distribution grid for delivering high-speed connections to the Internet. On April 23, 2003 the FCC announced a Notice of Inquiry seeking public comment on what it calls "Broadband over Power Line" (BPL): the use of existing electrical power lines as a transmission medium to provide high-speed communications capabilities, including Internet and broadband services, to both urban and rural areas by coupling radio-frequency energy onto the power line.

Btu (British thermal unit): The amount of thermal energy necessary to raise the temperature of one pound of water by one degree Fahrenheit at sea level.

bundled cabling: A type of cable available from vendors described in addendum 3 to the ANSI/TIA/EIA 568A standard that makes installation of Grade 2 structured wiring easier. The addendum describes four-pair cable assemblies that are not covered by an overall sheath (as specified for hybrid cables) but by any binding method, such as speed-wrap or cable ties. A standard bundled cable for Grade 2 consists of two Category 5e cables and two RG-6 quad-shield cables.

byte: A group of bits generally accepted in the computer industry as 8 bits (but 9 bits on 36-bit machines). Some older computer architectures used "byte" for quantities of 6 or 7 bits but these usages are now obsolete. An octet always contains 8 bits, and this term is used to avoid confusion with the term "byte."

C

cable tester: A type of network cable test equipment used to check cables for opens, shorts, and continuity.

call waiting: A local exchange telephone service option that notifies a subscriber conducting a call that one or more other phones are trying to make a connection. With call waiting service, the second call is announced by a soft beep. The user can ask the first caller to wait while she answers the second call. She then has the option to return to the first call or disconnect from the first to talk to the second caller.

caller ID: A local exchange telephone system service that identifies the number of the caller to the receiving (called) party. This service is one of many listed under the title of caller line identification or calling line identity (CLI) services. Features such as caller display and call return use a feature of modern telephone networks that transmits the number of the caller as each call is set up. Called customers using these CLI services have access to this information. With caller display, a customer can see the number on his phone before answering the call.

Category 5: A type of UTP communications cable described in ANSI/TIA/EIA 568A that has four twisted pairs of wires. Category 5 cable is tested to 100 MHz. Category 5 cable was previously used in residential structured wiring and 100Base-T Ethernet LAN installations. It has been replaced by Category 5e as the preferred structured wiring cable for new installations, which provides improved performance at approximately the same cost.

Category 5e: An enhanced version of Category 5 cable described in specification ANSI/TIA/EIA 568A addendum 5. Category 5e cable was officially ratified as a standard on December 13, 1999, and updated with the release of ANSI/TIA/EIA-568B.1-2000. Although it has a rated bandwidth of 100 MHz, it has improved specifications for near-end crosstalk (NEXT), power-sum equal-level far-end crosstalk (PSELFEXT), and attenuation. The specification also enforces several attributes that were optional in the original Category 5 specification. Category 5e cable is usually tested to a bandwidth of 350 MHz, despite its 100 MHz specified bandwidth.

Category 6: A type of data communications cable that complies with the transmission requirements in the TIA/EIA 568B.2-1 Commercial Building Telecommunications Category Standard. Cable carrying the Category 6 rating has a rated bandwidth of 250 MHz.

Category 7: A type of high-performance data communications twisted-pair cable that uses a braided shield surrounding all four foil-shielded pairs to reduce noise and interference. Category 7 cable, connecting hardware, and patch cords are rated to a maximum frequency bandwidth of 600 MHz.

Cathode-Ray Tube (CRT): A type of vacuum tube in which images are produced when an electron beam strikes a phosphorescent surface. CRTs are used as the primary video display device in computer monitors and television receivers.

CCTV: Abbreviation for closed circuit television.

CEBus (Consumer Electronics Bus): CEBus is a communications protocol developed by the Electronic Industries Alliance (EIA) as the ANSI/EIA-600 standard. CEBus is designed to control devices on a power line, but it also works on other media. The CEBus standard involves device addresses that are set in hardware at the factory, and includes four billion possibilities.

cipher locks: A secure access control system for doors, using personal access codes known by the user. Cipher locks operate by unlocking a door equipped with magnetic locks, when the correct programmed code is entered by the user on the cipher lock keypad.

cladding: The portion of a fiber-optic cable encircling the core material.

composite video: A single video signal that contains luminance, color, and synchronization information. NTSC, PAL, and SECAM are all examples of composite video systems.

condenser: A heat exchanger (outdoor coil) assembly used with residential air conditioning or heat pump systems. As an integral component of a heat pump or a central air conditioning system, it is located in a vented outdoor casing where a large fan passes air over the coils to exhaust heat. It is the transfer point where heat is transferred from inside the home to the outside air.

conduit: A type of enclosure for wiring, installed in hazardous locations such as potentially explosive air environments, underground wiring, or outdoor and moist locations. It is available in a variety of types for many environments. They range from rigid metal conduit, to intermediate metal conduit, to thin-walled conduit called electrical metallic tubing (EMT), to nonmetallic types called polyvinyl chloride (PVC).

copper communications riser cable (CMR): A fire rating for riser cable established by the National Electrical Code (NEC) ANSI/NFPA 70 standard. The NEC assigns fire ratings to communications transmission cables and discusses suitable applications for each cable type. Copper communications riser cable is suitable for installation in vertical riser shafts that run from floor to floor. It cannot be used in any environmental air spaces, such as air conditioning ducts, unless the local code allows it. CMR cable might also be labeled as flame test (FT)-4, and is not as expensive as plenum cable.

cross-connect: A facility enabling the termination of cable elements and their interconnection, and/or cross-connection, primarily by means of a patch cord or jumper.

crossover cable: A cable used to directly connect two computers, or two hubs, in which the receive and transmit wires of a twisted-pair cable have been transposed at each end of the cable connector to perform the crossover function normally performed by a single hub.

crossover networks: A device used with multiple speaker systems to separate portions of the audio spectrum, and routing them to individual speakers for optimum performance.

D

daisy chain: A type of residential wiring design used in home lighting systems, in which load requirements for individual lighting systems do not require a dedicated home-run connection to the main power distribution panel. Some lighting systems include multiple light fixtures controlled with wall switches that share a 15-ampere daisy-chain service from the main power distribution panel.

digital: Circuitry or telecommunications transmission systems in which information-carrying signals are restricted to either of two states, corresponding to logic 1 or 0.

Digital Subscriber Line (DSL): A high-speed Internet access service offered by local telephone companies using regular telephone lines. DSL can move data over the phone lines at speeds up to 6 Mbps, or 140 times faster than the fastest analog modems (56,000 bits per second). In addition to its very high speed, DSL has many benefits over analog connections. Unlike dial-up connections that require analog modems to dial in to the Internet service provider every time, DSL connections are always on. Another benefit is that the user can talk on the telephone at the same time he is accessing the Internet.

Digital Versatile Disc (DVD): A high-volume storage media with the same physical appearance and dimensions as a compact disc. Initially called digital video disks by the creators, DVDs were named digital versatile disks in 1996. DVDs are standardized with five physical formats and four media storage versions.

DVD-ROM is a high-capacity data storage medium. DVD-Video is a digital storage medium for feature-length motion pictures, and DVD-Audio is an audio-only storage format similar to CD-Audio. DVD-R offers a write-once, read-many storage format akin to CD-R. DVD-RAM was the first rewritable (erasable) flavor of DVD to come to market and has subsequently found competition in the rival DVD-RW and DVD+RW formats. With the same overall size as a standard 120-mm diameter, 1.2-mm thick CD, DVD discs provide up to 17 GB of storage with transfer rates that are higher than those of CD-ROMs, and access times similar to CD-ROMs. DVDs come in four versions. DVD-5 is a single-sided, single-layered disc, boosting capacity sevenfold to 4.7 GB. DVD-9 is a single-sided, double-layered disc offering 8.5 GB, whereas DVD-10 is a 9.4 GB dual-sided, single-layered disc. Finally, DVD-18 increases capacity to a huge 17 GB on a dual-sided, dual-layered disc. DVDs are similar in appearance to CDs — they are both 4 ¾ inches in diameter. DVD tracks, however, are closer together and the compression ratio is much higher. The result is that DVD discs can hold 4.7 GB per side compared to approximately 660 MB for a CD.

DIN: A common type of connector used for video and computer interfaces. DIN is an acronym for *Deutsches Insitut für Normung eV*, a standards-setting organization for Germany.

Dolby Digital (formerly known as AC-3): A 5.1-channel surround sound format, and the standard for DVD sound tracks. Dolby Digital features five discrete Dolby Surround channels (front center, front left, front right, surround left, surround right; giving it the "5" designation) of full-frequency sound, plus a sixth channel for low-frequency effects (LFE).

Dual-Tone MultiFrequency (DTMF): A method of tone dialing using a 12-button keypad with dual tones representing each of the 12 buttons.

Dynamic Host Configuration Protocol (DHCP): A network protocol that enables individual computers on an IP network to obtain their IP address configurations from a DHCP server. This reduces the work necessary to administer a large IP network by an administrator. The DHCP server allocates an IP address to the PC from one of the scopes (the pools of addresses) it has available.

dynamic speaker: An audio reproduction device, also referred to as a cone speaker, in which the audio signal is applied to a voice coil that is part of the moving system.

E

ElectroMagnetic Interference (EMI): A type of disturbance occurring in communications circuits when unwanted electrical currents are induced into the wiring or components of an electrical system. EMI can also degrade the performance of wireless communications channels.

electromagnetic lock: A type of lock used to control access through a door where maximum security is required. It consists of an electromagnet mounted in the door frame and a matching plate installed in the door. The electromagnet can be energized from a switch by the person controlling the access. When energized, the electromagnet holds the plate with a strong magnetic field. The lock has no moving parts. The amount of pressure an electromagnetic lock can withstand is somewhat determined by its integrity rating. The Builders Hardware Manufacturers Association (BHMA) together with the American National Standards Institute (ANSI) has determined three integrity grades for electromagnetic locks. A Grade 3 integrity rating provides a locking strength of 500−900 lb. An integrity rating of Grade 2 means that the electromagnetic lock must be strong enough to withstand a pressure of 1,000−1,400 lb. And an integrity rating of Grade 1 means that the electromagnetic lock is designed to withstand pressures of 1,500−2,700 lb.

electromechanical controller: A type of irrigation system controller used to operate the valve stations at prescribed times. Electromechanical controllers use clocks in addition to stored irrigation programs. They are respected as a reliable solution for simple irrigation systems. They have few sophisticated electronic components and are driven by electric motors and gears. Turning dials or flipping switches programs the controller to select watering times, how long each zone will be watered, and on which days the watering will occur.

electrostatic speaker: A special type of speaker designed to avoid the problems inherent in cone speakers. They use a graphite-coated plastic membrane suspended between two perforated metal sheets. A high voltage of several thousand volts is applied to the membrane. The input signal is also raised to a high voltage by a transformer, and is applied to the perforated metal sheets.

encapsulation: A communications process in which a packet or frame is enclosed within another packet for the purpose of hiding the header of the encapsulated packet during transit through a network. At the destination, the receiving node recognizes the encapsulation header and recovers the original enclosed packet.

Ethernet: A popular local area network technology that is standardized by the IEEE as the 802.3 Carrier Sense Multiple Access with Collision Detection (CSMA/CD) standard. Ethernet was originally developed by two engineers at the Xerox Corporation and then developed further by a partnership of Xerox, DEC, and Intel. An Ethernet LAN typically uses coaxial cable, or special grades of UTP cable, as the transmission media.

European Broadcasting Union (EBU): A professional society that helps establish standards in the audio and broadcast industry in Europe, as well as other nations.

evaporator: A component of a heat pump or central air conditioning system located inside the home. It is also known as an *indoor coil* and functions as the heat transfer point for warming or cooling indoor air. It consists of a series of pipes connected to a furnace, or air handler, which blows the indoor air across the evaporator coil, causing the coil to absorb heat from the air. The cooled air is then delivered to the house through the ducting.

F

Fast Ethernet: The common name for the IEEE 802.3u Ethernet 100Base-T specification that transmits information at a wire speed of 100 Mbps.

fax: An abbreviation for facsimile machines used in home offices or businesses to send and receive documents using standard telephone lines, or dedicated digital lines. Faxing is a less expensive alternative to overnight package delivery, and can respond to situations in which a document is needed the same day for an urgent business transaction.

firewall: A system designed to prevent unauthorized access to, or from, a private network. Firewalls can be implemented in hardware or software, or a combination of both. Normally, a firewall is deployed between a trusted, protected private network and a public network. For example, the trusted network might be a corporate network and the public network might be the Internet.

FireWire: The product name for Apple Computer's version of the standard IEEE 1394 High Performance Serial Bus. It is used for connecting peripheral devices to personal computers. FireWire provides a single plug-and-socket connection on which up to 63 devices can be attached, with data transfer speeds of up to 400 Mbps.

fixed wireless access systems: Fixed wireless includes facilities for broadband service originating from ground-based transmission sites as well as satellites. Similar to satellite service, it is an alternative to technologies that use wired connections such as telephone and coaxial cable networks. The two FCC-licensed commercial services available in the fixed wireless service are called multichannel multipoint distribution service (MMDS), and local multipoint distribution service (LMDS). The multichannel multipoint distribution service can provide Internet-access downlinks over a distance of about 50 km from a central-transmitter site, using radio frequencies in the 2.1 and 2.6 GHz band. LMDS services use radio frequencies above 10 GHz—frequencies referred to as millimeter waves, at 28 GHz and 38 GHz. However, these are fiber-optic network replacement technologies intended for higher-density urban areas.

flat-panel display: A common name used to describe plasma or LCD displays. Flat-panel displays get their name from the type of construction that sets them apart from CRT displays. They are suitable for laptop computers and provide portable computing, with the technology to make them truly portable and lightweight. Flat screens are a special type of CRT display.

frequency: Within the context of wireless communications terminology, the number of cycles or reversals of an electromagnetic field that occur in one second. Frequency also expresses the quantity of units of information transmitted in a given time period, expressed in hertz.

front projection: A method of viewing video images that utilizes a video projector and a separate pull-down screen to show the projected image, similar to a movie screen.

G

gateway: A special home network interface device, similar to a router, that connects a single Internet access line such as DSL, cable, or telephone to the home computer network. The residential gateway serves as the communication hub for the entire suite of home networking devices deployed throughout the home. Gateways provide integrators with two functions — one is to connect the home network to the Internet, and the other is to act as a central communications interface resource for all in-home digital devices, ranging from lighting, security, and HVAC, to phones, home entertainment components, and personal digital assistants. Various types of residential gateways are available, based on the gradients of functionality required. Most provide capability for advanced data, voice, and video routing within the home. Most gateway products also include a firewall. The gateway is normally located near the distribution panel to simplify the connections from the gateway, to the home network patch panel, and the incoming Internet access port.

G.DMT (G.Discrete Multitone): A type of ADSL service defined in ITU Recommendation 992.1. The main difference between G.Lite and G.DMT is bandwidth. Sometimes called full-rate ADSL, the G.DMT variety can download data at up to 8 Mbps, and send data upstream at up to 1.5 Mbps

G.Lite: A type of Universal DSL described in ITU Recommendation G.992.2. Sometimes called *splitterless DSL*, it is a form of ADSL that does not require a splitter installation at the subscriber location, but at the expense of lower data rates. G.Lite can be used with a distributed splitter configuration, but with added complexity and limited benefit. It represents the most consumer-friendly version of DSL. G.Lite equipment and service costs less than other varieties, and it reportedly has a do-it-yourself installation. G.Lite supports a maximum of 1544 Kbps downstream, and 384 Kbps upstream.

Gigabit Ethernet: A term used to describe the IEEE 802.3z 1000Base-X Ethernet specification. Gigabit Ethernet, a transmission technology based on the Ethernet frame format and protocol used in LANs, provides a data rate of 1 billion bits per second (1 gigabit). Gigabit Ethernet is carried primarily on optical fiber cable (with very short distances possible on copper media).

GPM: Abbreviation for "gallons per minute."

Grade 1: The grade level described in ANSI/TIA/EIA 570A (Residential Telecommunications Cabling Standard) that provides a generic cabling system that meets the minimum requirements for telecommunication services. It specifies a minimum of one twisted-pair cable and associated connecting hardware (Category 3) and one coaxial cable (75-ohm) configured in a star topology. This grade provides for telephone, TV (digital or analog), satellite, CATV, and low-speed data services.

Grade 2: The grade level described in ANSI/TIA/EIA 570A (Residential Telecommunications Cabling Standard) in which each cabled location outlet requires two Category 5 UTP cables and two 75-ohm coaxial cables plus, as an option, optical fiber cabling configured in a star topology. This grade provides a broader range of residential services than Grade 1.

graphic equalizer: An audio unit that includes a set of filters, each with a fixed center frequency that cannot be changed. The only control available is the amount of boost or cut in each frequency band. This boost or cut is usually controlled with sliders. This interface is intuitive because the frequency response of the equalizer resembles the positions of the sliders themselves.

Ground Fault Circuit Interrupter (GFCI): A type of electrical outlet used to protect people from electrical shock due to faulty extension cords, appliances, or tools. A GFCI detects any minor imbalance in the current flowing in the hot and neutral conductors that might be caused by a ground fault in the connected appliance, and disconnects the outlet automatically.

H

heat pump: A special type of HVAC system that can heat a home during the winter and cool it during the summer. By definition a heat pump is any device that accepts heat at one or more temperatures, and rejects heat at a higher temperature.

High-Definition Television (HDTV): An advanced television broadcast standard adopted by the ATSC that provides higher resolution, a 16:9 aspect ratio, and 5.1 surround sound channels.

Home Audio/Video Interoperability (HAVi): A vendor-neutral audio/video standard aimed specifically at the home entertainment environment. HAVi allows different home entertainment and communication devices (such as VCRs, TVs, stereos, security systems, and video monitors) to be networked together and controlled from one primary device, such as a PC.

HomePNA (HPNA): The HomePNA abbreviation stands for Home Phone Network Alliance. It is a standard based on a process that enables voice and data transmissions to share the available bandwidth of the telephone cabling in a home, without mutual interference. It uses frequency division multiplexing (FDM) techniques to divide this bandwidth of frequencies into several communications channels. The acronyms for phone-line networking and power-line networking standards organizations can be confusing. The standard adopted for using the phone lines for networking is derived from the standards group that developed it (Home Phone Network Alliance). The published standard is often referenced as HPNA 2.0. (See also "HPLA.")

HomeRF: A wireless LAN home technology standard supported by the HomeRF Working Group. It is not compatible or interoperable with the IEEE 802.11 family of wireless LANs. The founding members of the HomeRF consortium include, among others, Microsoft, Intel, HP, Motorola, and Compaq.

home-run wiring: A method used in modern automated home design in which all the wiring is routed from a central point in the house, usually in a wiring closet. At the central point, the wires terminate at various types of patch panels or cross-connect terminals. This point becomes the interface between home wiring and external telecommunications wiring. Home-run wiring is the basic design concept used in structured wiring.

horizontal resolution: Chrominance and luminance resolution (detail) expressed horizontally across a picture tube. This is usually expressed as a number of black-to-white transitions, or lines, that can be differentiated. Horizontal resolution is limited by the bandwidth of the video signal, or equipment.

HPLA: An acronym for a standard called the HomePlug Powerline Alliance. The name is derived from a group of companies that developed the standard for power-line networking. The standard is referenced as HomePlug 1.0 or sometimes HPLA 1.0.

Hybrid Fiber Coaxial (HFC): A network that includes both fiber-optic cable and coaxial cable as the transmission media. HFC networks are used by the cable TV industry for the distribution of digital video channels and high-speed Internet access services.

Hypertext Markup Language (HTML): Authoring software used on the World Wide Web. HTML is basically ASCII text surrounded by HTML commands.

HVAC: Abbreviation for heating, ventilation, and air conditioning.

I

IEEE 802 standards: A series of standards approved by the IEEE, as follows:

- 802.1 (High-level Interface)
- 802.2 (Logical Link Interface)
- 802.3 (CSMA/CD)
- 802.4 (Token Bus)
- 802.5 (Token Ring)
- 802.6 (Metropolitan Area Networks)
- 802.7 (Broadband LANs)
- 802.8 (Fiber Optics)
- 802.9 (Integrated Voice Data)
- 802.10 (LAN Security)
- 802.11 (Wireless Networks)
- 802.12 (Demand Priority Access (100VG AnyLAN))
- 802.13 (Not Used)
- 802.14 (Cable TV-based Broadband)
- 802.15 (Wireless Personal Area Network)
- 802.16 (Broadband Packet)

IEEE-1394: Also known as FireWire, IEEE-1394 is a very fast external serial bus standard that supports data transfer rates of up to 400 Mbps. Products supporting the IEEE-1394 standard have different names, depending on the company. Apple, which originally developed the technology, uses the trademarked name FireWire. Other companies use other names, such as i.Link (Sony) and Lynx (TI), to describe their IEEE-1394 standard products. A single 1394 port can connect up to 63 external devices.

Insulation Displacement Connection (IDC): The recommended method of telecommunications copper wire termination recognized by the ANSI/TIA/EIA 568A standard. These termination devices are the popular choice for home network star topology connection points. Commonly called punchdown connections, these connections require the use of a small punchdown tool to properly secure the cable to terminal block. Punchdown connections remove or displace the conductor's insulation as it is seated in the connector. During termination, the cable is pressed between two edges of a metal clip, which displaces the insulation and exposes the copper conductor. This ensures a solid connection between the copper conductor and the terminating clip.

Integrated Services Digital Network (ISDN): A type of broadband, all-digital access service offered by local exchange carriers. ISDN provides circuit-switched access to the public network for voice, data, and video transmission. Basic rate interface (BRI) and primary rate interface (PRI) are two types of ISDN service.

interlaced scanning: A type of scan used in television broadcasting in which half the screen image is developed on successive frames of odd and even lines.

Intermodulation Distortion (IM): Intermodulation distortion is a measured performance feature of an amplifier that is listed in the specifications. It is caused when two or more frequencies become mixed in a nonlinear device. New frequencies that are not part of the source information are generated and produce undesirable effects in the amplifier output.

Internet: A worldwide collection of networks that make up an international information source, where almost any client can access any server. The Internet was initially a U.S. government network called the ARPAnet, which was named from the developing organization, the Advanced Research Projects Agency.

Internet Assigned Number Authority (IANA): An organization that previously performed services for IP address assignments and related engineering services, which are now performed by ICANN.

Internet Corporation for Assigned Names and Numbers (ICANN): A nonprofit corporation that was formed to assume responsibility for the IP address space allocation, protocol parameter assignment, domain name system management, and root server system management functions previously performed under U.S. government contract by IANA, and other entities.

Internet Engineering Task Force (IETF): An international community of network designers, operators, vendors, and researchers concerned with the evolution of the Internet architecture, and the operation of the Internet. The actual technical work of the IETF is done in its working groups.

Internet Protocol (IP): A primary protocol included in the TCP/IP suite used for encapsulating other protocols, such as TCP and UDP. The IP header includes the source and destination addresses of the packet, and other routing information.

Internet Service Provider (ISP): A company that provides individuals, and other companies, access to the Internet and other related services such as Web site building, and virtual hosting. An ISP has the equipment and the telecommunication line access required to have a point-of-presence on the Internet for the geographic area served.

ionization detector: A type of sensor used in home smoke detectors. It functions by detecting the disturbance caused by smoke in a small ionization chamber.

IrDA: IrDA stands for Infrared Data Association. It is an international organization that creates and promotes interoperable infrared data interconnection standards for wireless networks. The name is associated with products that conform to the interconnection standard.

ISO: A name used to identify the International Organization for Standardization. Because "International Organization for Standardization" would have different abbreviations in different languages ("IOS" in English, "OIN" in French for Organisation Internationale de Normalisation), the organization decided to use a word derived from the Greek "isos," meaning "equal." Therefore, whatever the country, whatever the language, the short form of the organization's name is always ISO. It is often used incorrectly as an acronym for International Standards Organization.

J

Jini: (pronounced GEE-nee) A Java-based standard from Sun Microsystems, which enables a plug-and-play network based on TCP/IP. Jini makes home networks dynamic by allowing devices and services to enter and leave the home network without complicated setup procedures. Jini shields the user from the installation, configuration, and setup process. Because Jini technology is also platform-independent, devices are no longer limited by specific brands of software, processors, device drivers, or traditional networking protocols. Jini is a direct competitor to HAVi, as it also defines a protocol for a plug-and-play networks.

junction (electrical) box: A square, octagonal, or rectangular plastic or metal box that fastens to framing, and houses wires, receptacles, and switches.

K

key system: A type of privately owned telephone system used by small business firms and home offices. It is less expensive than a PBX, and is designed for organizations that need as few as 3 or 4 phones, to as many as 250 phones. Key system phones have buttons that select outside lines, and can call other phones internally.

Most key systems include a controller-type cabinet called a key service unit (KSU), and a set of proprietary telephones. The KSU holds all the system's switching components, the system power supply, outside lines, and internal station cards.

kilowatt-hour meter: A device installed by the local electrical power utility to measure the power consumed in a home. The meter records the cumulative power consumed in kilowatt-hours. A single-phase watt-hour meter is essentially an induction motor, whose speed is directly proportional to the voltage applied, and the amount of current flowing through it. The phase displacement of the current, as well as the magnitude of the current, are automatically taken into account by the meter. In other words, the power factor influences the meter's speed and the moving element (disk) rotates with a speed proportional to true power. The read-out dials are simply a means for counting the motor revolutions, and by proper gearing, are arranged to read directly in kilowatt-hours.

L

LAN: Acronym for local area network. Also used to describe wireless local area networks (WLANs).

laserdisc: An older analog video optical disc format, first introduced in 1974. Each side of a disc could hold about an hour of video, so feature films required a pause in the middle to flip the disc over. Some films are on two discs. The introduction of DVDs made laserdiscs obsolete for the video market.

Liquid Crystal Display (LCD): A type of flat display technology used in laptop computer design, calculators, PDAs, and the majority of flat-screen displays. LCD displays are available as active matrix, dual-scan, or passive matrix display types. An LCD uses the fact that certain organic molecules (liquid crystals) can be reoriented by an electric field. As these materials are optically active, their natural twisted structures can be used to alter the polarization of light on a flat screen.

LNB: A component used in satellite receiving antenna feed systems. LNB is an acronym for Low Noise Block-down converter (so called because it converts a whole band or "block" of frequencies to a lower band).

load: The maximum amount of electrical current (in amperes) required for all the electrical devices in a home. Load analysis is usually performed by an electrical contractor for new home or remodeling construction. Load is also used to determine the power consumed by household electrical appliances. The actual power used by the load is called *true* power, or just *power*, and is measured in watts. (Even though watts = volts x amps, apparent power is measured in VA to differentiate it from true power.)

Local Exchange Carrier (LEC): Any public telephone company in the United States that provides local telephone service. Some of the largest LECs are the Bell operating companies (BOCs), which were grouped into holding companies known collectively as the regional Bell operating companies (RBOCs), when the Bell System was broken up by a 1983 consent decree. LEC companies are also sometimes referred to as *telcos* or *telephone companies*. A *local exchange* is the local central office of an LEC. When you pick up a phone to make a call, the dial tone you hear is coming from the central office of your local exchange carrier. Telephone lines from homes and businesses terminate at a local exchange facility. Local exchanges connect to other local exchanges within a local access and transport area (LATA).

LonWorks: A multipurpose home networking protocol developed by the Echelon Corporation that can be implemented over any medium, including power lines, twisted-pair cable, wireless (RF), infrared (IR), coaxial cable, and fiber optics. It has no central controller but uses a system of intelligent nodes that communicate with each other. The LonWorks protocol was published as ANSI/TIA/EIA-709 in 1998.

low-voltage wiring: All residential cabling that is not associated with 240V/120V high-voltage alternating current (AC) primary power wiring used in all U.S. residential construction. Although no exact definition of low-voltage exists in residential cabling standards, it is generally accepted to include those circuits that are connected to low-energy current-limited sources, as described in Article 725 of the NEC. Low-voltage wiring includes all wiring for audio/video components, network data cabling, security sensors, phone lines, telecommunications systems, and low-voltage landscape lighting. This encompasses all residential wiring, with the exception of 120-volt AC wiring.

lux: The amount of light required to obtain a reasonable CCTV video camera image. One lux is approximately the light from one candle measured from one meter. Typical camera ratings range between 0.5 and 1.0 lux.

M

magnetic stripe card: A type of security card used to control access to restricted areas, or residential gated communities. It consists of a plastic card with a narrow strip of magnetic material fused to the back. Data stored on the strip as narrow bars form the basis of a binary code, which is used by the access control system reader.

MC cable: Cable with a flexible metal covering. MC cable is composed of THHN soft-drawn copper wire conductors, and an insulated grounding conductor. It is suitable for branch, feeder, and service power distribution in commercial and industrial applications, as well as in multifamily buildings, theaters, and other populated structures. The acronym *THHN* refers to heat-resistant thermoplastic (90° C) for dry locations. This is a reference to the type of insulation on the wiring, coded per the NEC.

middleware: Connectivity software with enabling services, that allows multiple processes running on home computers to interact across a network. The main requirements for middleware are independence of any specific networking technology, interoperability between devices of different manufacturers, common application programming interfaces (APIs) that allow application development and service provision by third parties, and a common user interface framework that supports the specific requirements of the residential users, and their environment.

motorized damper: A component of an HVAC system that regulates the amount of warm or cold air that enters a room, or an area.

Moving Picture Experts Group (MPEG): A set of video and audio compression standards. The MPEG format employs compression algorithms to remove redundant picture information from successive video scenes. The MPEG methodology compresses only key objects within a frame, every 15th frame. Between the key frames, only the information that changes from frame to frame is recorded. The MPEG standard includes compression specifications for both video and audio signals.

MMDS (Multichannel Multipoint Distribution Service): A type of fixed, wireless communication service that uses FCC-licensed radio spectrum. Although originally developed as a one-way wireless service to deliver television service as a competitor to cable TV companies, MMDS has been permitted, through relaxed federal telecommunications legislative action, to provide broadband Internet service to business and residential customers.

multiplexer: An electronic, high-speed switching device that combines information contained on several digital video, or data channels, into a single high-speed channel. Multiplexers are used with home surveillance systems for recording several channels simultaneously.

N

National Electrical Code (NEC): A document sponsored by the National Fire Protection Association (NFPA). The National Electrical Code covers the requirements for electric conductors, and equipment installed within, or on, public and private buildings, or other structures. This includes mobile homes and recreational vehicles; floating buildings; and other premises such as yards, carnivals, parking and other lots, and industrial substations. It also covers conductors that connect the installations to a supply of electricity, and other outside conductors and equipment on the premises, including optical fiber cable. It covers buildings used by electric utilities, such as office buildings; warehouses; garages; machine shops; and recreational buildings that are not an integral part of a generating plant, substation, or control center.

Most states require a permit and an inspection for electrical work. Even though following the NEC cannot guarantee safe electrical installations, it is the best guide available. However, every state might differ slightly in its requirements for inspection and code compliance.

National Electrical Contractors Association: An international building trades organization that includes over 70,000 electrical contracting firms. NECA has 120 U.S. chapters, in addition to others in countries around the world.

National Electrical Manufacturers Association (NEMA): An organization of manufacturers of electrical equipment, including, but not limited to, wiring devices, wire and cable, conduit, load centers, pressure wire connectors, circuit breakers, and fuses. NEMA is the voice of the electrical industry, and through it, standards for electrical products are formulated. Generally, these standards promote interchangeability between products of one manufacturer, and similar products made by another manufacturer. In some cases, standards relating to product performance are also formulated by NEMA, but these are the exception rather than the rule.

National Fire Protection Association (NFPA): A U.S. standards organization that develops, publishes, and disseminates timely consensus codes and standards intended to minimize the possibility and effects of fire and other hazards. Virtually every building, process, service, design, and installation in society is affected by NFPA documents.

National Television System Committee (NTSC): The organization that developed the U.S. broadcast television standards. NTSC has also become the name of this group of standards. Other countries, such as Canada, Mexico, and Japan, have also adopted the NTSC broadcast standard.

Network Address Translation (NAT): A software feature used by gateways and routers to interconnect a private network that uses unregistered IP addresses with a global IP network that uses a limited number of registered IP addresses. NAT hides private IP addresses from the Internet, while still permitting the computers on that network to access the Internet.

Network Interface Card (NIC): A network interface card is a printed circuit board, with a connector that plugs into a PC expansion slot. A NIC contains a transceiver for sending and receiving data frames on a network, as well as the Data-Link layer hardware needed to format the sending bits, and to decipher received frames.

NH cable: A type of cable used in electrical wiring that is halogen-free (NH stands for nonhalogen). The noncorrosive material chosen enables this cable to be used for fixed installations in public buildings, and in governmental installations, where halogen-free products are demanded. The jacket is made of flame-retardant polyolefin material.

NM cable: Abbreviation for nonmetallic (NM) cable, the most widely used electrical power cable for indoor wiring. Romex is a brand name for a type of plastic insulated wire. It is often called nonmetallic sheath; however, the formal code name is NM cable. This type of wire is suitable for stud walls, on the sides of joists, and other areas that are not subject to mechanical damage or excessive heat. Newer homes are wired almost exclusively with NM wire.

O

octet: Eight bits in an IP address. An octet is thus an eight-bit byte. Since a byte is not eight bits in all computer systems, *octet* is used as a precise reference to eight bits. Technically, an octet represents any 8-bit quantity. By definition, an octet ranges in mathematical value from zero to 255. Typically, an octet is also a byte, but the term "octet" came into existence because historically, some computer systems did not represent a byte as eight bits.

optoelectronic: Describes any device that functions as an electrical-to-optical or optical-to-electrical transducer, or an instrument that uses such a device in its operation.

oxidation reduction potential (ORP): A measure of a swimming pool water's overall capability to eliminate wastes. High oxidation is present when pollution is low, and the water is of high quality. ORP is rapidly becoming the standard means of testing and regulating pool water sanitation.

P

packet: A unit of information formatted for transmission over a network. A packet contains control and header information that corresponds to the OSI Network layer, or layer 3. *Packet* is often used interchangeably (and incorrectly) with *frame*.

passive infrared (PIR): A type of motion detector sensor used in home security systems. The term *infrared* comes from the sensor's capability to see areas of the optical spectrum outside the range of light frequencies discernible by human vision. PIR motion detectors operate on the principle that heat, or infrared emissions, emanating from a human entering an area, will disturb an otherwise stable infrared background and trigger an alarm. Any area that does not change abruptly, over time, has a normal infrared background temperature. A PIR is referred to as passive because it does not release any energy of its own, but instead, causes an alarm when the surrounding infrared background environment changes abruptly.

peer network: A peer network (or peer-to-peer network), as the name suggests, is where all computers on a LAN are peers, and have equal status concerning sharing rights and capabilities. Any individual machine can share data or peripherals with any other machine on this type of network.

Personal Communications Service (PCS): A second-generation U.S. digital cellular telephone service established by the Federal Communications Commission. PCS services use the 1900 MHz band. All the licenses for the use of the PCS frequency bands by various service providers were awarded through public auctions for services in the major metropolitan areas. The FCC has also allocated spectrum for unlicensed PCS services in the 1910 MHz−1930 MHz band. The unlicensed PCS service accommodates a wide range of services for small in-building areas, such as data networking within office buildings, wireless private branch exchanges (PBXs), personal digital assistants, laptop computers, portable facsimile machines, wireless replacements for portions of the wire-line telephone network, and other types of short-range communications.

Personal Video Recorder (PVR): A video recorder similar to a Video Cassette Recorder (VCR), except it uses a hard disk instead of a magnetic tape as the recording media. PVRs are designed to record live TV broadcast streams "on the fly" and allow the viewer to pause a program, take a break, and resume watching a program, even though the broadcast has continued past that point.

Phase Alternate Line (PAL): A color television broadcast standard used in Europe. Broadcasts in NTSC format can have problems with color hues varying in unwanted ways because of phase shifts in the color subcarrier. PAL fixed that by reversing the phase on every other line, hence the name. Because TV came later to Europe than to the United States, this superior format was adopted there instead of NTSC.

phone-line splitter: A device installed on a telephone line that is connected to both high-speed broadband digital service and analog voice devices, such as telephones and fax machines. The splitter routes the broadband and analog signals on the telephone line to the correct device. Most splitters must be installed by the telephone company; however, some can be installed by the customer.

photoelectric switch: A special type of switch used for controlling lights, depending on the light intensity (lux value). This type of switch contains a special sensor that operates when the light intensity, or darkness, reaches a certain value. Photoelectric switches are typically used outdoors for landscape and security lighting. Most photoelectric switches are integrated in a single fixture that includes the sensor and lamp, and have adjustments to turn off lights after a number of hours, where it is unnecessary to have lighting from dusk to dawn.

Plain Old Telephone Service (POTS): A name often used to describe the analog public switched telephone network.

Plasma Display Panel (PDP): A type of advanced large-screen display technology that uses hundreds of thousands of tiny cells (pixels) containing minute amounts of an inert gas, sandwiched between two sheets of glass. Electrodes are placed in pairs on the inner side of the front plate. When electrically charged, they produce an ultraviolet beam, which activates the phosphorous coating of the cell, and transmits light through the glass surface. The brightness of the emission depends on the current passing through the ionized gas. Large and brilliant color images can be produced with different colors of phosphors. Plasma displays provide a totally flat screen design with excellent off-center viewing capability, and superior resolution.

plenum: A compartment or chamber to which one or more air ducts are connected, and that forms part of the air distribution system.

Point-to-Point Tunneling Protocol (PPTP): A network protocol that encapsulates Point-to-Point Protocol (PPP) packets into IP datagrams for transmission over the Internet, or other public TCP/IP-based networks. PPTP can also be used in private intranets.

polyethylene: A plastic used for manufacturing irrigation tubing.

pop-up sprinkler: A popular type of sprinkler head used for home lawn and garden irrigation systems. It is installed below the ground, and the sprinkler head remains out of sight when it is inactive. When the sprinkler system is turned on, a small portion of the head emerges above the surface to disperse water to the irrigation area. Spray head sprinkler pop-ups are designed to spray a small fixed area, whereas rotor types are normally used for large, unobstructed areas.

Power Line Carrier (PLC): The capability for the power-line wiring used in the electrical system of a home to carry a command signal for controlling lighting and appliances. The X-10 control protocol that uses PLC technology was invented to exploit this capability.

Pressure Vacuum Breaker (PVB): A type of backflow preventer used with irrigation systems.

Private Branch Exchange (PBX): A telephone switch owned by a private organization or home office user. The system is usually located at the owner's facility. The PBX provides phone services including internal calling and access to the public switched telephone network. It allows a small number of outside lines to be shared among all the people of the organization. Advanced PBX phone switches sometimes provide auto-attendant, voice-mail, and ACD (automatic call distribution) services for the organization.

programmable thermostat: A special type of thermostat that enables temperature adjustments to be scheduled according to the day of the week and the time of day. It controls both heating and air conditioning equipment with a single integrated control processor.

progressive scan: A high-definition TV (HDTV) format. The progressive scan system scans the total number of lines, 60 times per second. This means that you see the complete image displayed on your TV screen twice as often than in the interlaced scan method. This results in smoother motion in moving images, fewer motion artifacts, and no visible flicker. A progressive scan system with 720 lines of resolution is written as 720p. Progressive scan has been used for many years for computer display monitors. It is also employed for DVD players that play back DVD movies in the progressive scan mode, for viewing on advanced digital television receivers.

protocol: A formal set of rules that establishes the method of handling data transmissions in both wired, and wireless, networks.

psi: Abbreviation for "pounds-per-square-inch."

Public Safety Answering Point (PSAP): A location where emergency response dispatchers answer 911 calls. The dispatch operator at the PSAP obtains as much information as possible from the caller, and sends the appropriate help to the caller's location. The caller might be in distress and unable to provide location information. The caller's phone number, however, is available to the dispatch operator through a system called automatic number identification (ANI). ANI provides the receiver of a telephone call with the number of the calling phone. The method of providing this information is determined by the local exchange carrier.

pull station: A device, typically mounted on a wall, that is used to manually activate a fire alarm system.

punchdown tool: A tool used to connect wires to an insulation displacement connection (IDC) block. This type of termination is the recommended method of copper termination recognized by the ANSI/TIA/EIA 568A standard for UTP cable terminations. These termination devices are the popular choice for making home network star topology connection points. Commonly called punchdown connections, these connections require the use of a small punchdown tool to properly secure the cable to the terminal block. Punchdown connections remove or displace the conductor's insulation as it is seated in the connector. During termination, the tool presses the cable between two edges of a metal clip, which displaces the insulation and exposes the copper conductor. This ensures a solid connection between the copper conductor and the terminating clip.

R

Registered Jack (RJ): In the United States, telephone jacks are also known as *registered* jacks, sometimes described as RJ-XX. This series of telephone connection interfaces (receptacle and plug) is registered with the FCC. They originate from interfaces that were part of AT&T's Universal Service Order Code (USOC), and were adopted as part of FCC regulations (specifically Part 68, Subpart F, Section 68.502). The term *jack* sometimes means both receptacle and plug, and sometimes just the receptacle.

relay: An electrical switching device often used in low-voltage control circuits to switch higher voltages on and off, such as a motor or lighting circuit.

remote access: A type of telecommunications service that provides user access to a computer, or network, from a remote location. It's often used by employees on business travel, who need access to the corporate private network to read e-mail or to access shared files. Remote access is also a valuable tool for home network users who need to access automated home features from a remote location.

Remote Authentication Dial-In User Service (RADIUS): A client/server protocol that enables remote access servers to communicate with a central server on a LAN, to authenticate dial-in users and authorize their access to the requested system or service. RADIUS enables a company to maintain user profiles in a central database that all remote servers can share. It provides better security, allowing a company to set up a policy that can be applied at a single, administered network point.

reversing valve: A type of valve used by heat pumps to redirect the refrigerant flow. It enables the inside coil and outside coil to act as either an evaporator, or a condenser.

ribbon speaker: A special type of speaker where the input signal is applied to a foil ribbon suspended between magnets, or metal sheets. Ribbon speakers do not use a cone and voice coil; the input signal from the amplifier is applied to the foil ribbon. The varying electrical charge caused by the input signal forces it to be repelled from, or attracted to, the magnets, thereby moving air and producing sound. This design overcomes some of the deficiencies of cone speakers in the mid and upper ranges of the audio spectrum.

riser: A type of duct for holding wire and cable that connects one floor to another, usually in a commercial building, and penetrates fire-rated walls or floors. Cable installed in risers must meet specific fire standards published by the NFPA. This type of cable is referred to as riser cable.

RJ-11: A registered jack (RJ) connector used for terminating a telephone cable that connects the phone to a standard wall outlet. It is used by many home automation devices, and is the standard used for all residential phones lines.

RJ-31X connector: A type of registered jack used for attaching alarm systems to telephone lines. The connector takes control of the phone line during a security alarm condition.

RJ-45 connector: A type of registered jack that uses an eight-conductor connector, with a small locking pin. It's used to connect the network interface card in networked computers to a local area network.

rotor sprinkler: Rotor sprinkler heads are pop-up sprinklers that disperse water over a large circular area. Small rotors can cover radii of up to 50 feet, whereas large rotors cover radii up to 100 feet. The two types of rotor sprinklers are called impact, and gear-driven. They differ only in the mechanism used to move the head in a circular motion.

router: A component consisting of hardware and software that provides intelligent connections between networks. Routers operate at the Network layer (layer 3) of the OSI model, and are responsible for making decisions about which paths through a network the data packets will use.

S

S/PDIF: Sony/Philips digital interface format is a consumer version of the AES/EBU digital audio interconnection standard that is based on coaxial cable and RCA connectors.

S-Video: A video interface standard that is a compromise between component analog video and composite video, because it separates the luminance (brightness) and chrominance (color) information. It uses twin coaxial type cables integrated into a single cable and terminated with a DIN connector.

scene: The effect created by a lighting control system, when a number of lights dim and brighten to different intensities.

Screened Twisted Pair (ScTP) cabling: A new type of 100-ohm, twisted-pair cabling that evolved from the ANSI/TIA/EIA standard IS-729. ScTP cable types are being developed to meet new, more rigorous criteria for Category 6E and 7E cable.

SECAM (Sequential Electronique Couleur Avec Memoire): Translated as sequential color and memory, SECAM is a composite color television transmission system that potentially eliminates the need for both the color and hue controls. It has a 625-line vertical resolution and a frame rate of 25-frames-per-second. One of the color difference signals is transmitted on one line, and the other is transmitted on the second line. Memory is required to obtain both color difference signals for color. This standard is used in France, Eastern Europe, Africa, and Asia.

set points: Temperature settings used on residential programmable and communicating thermostats that indicate how high or low the user wants the temperature to be in a home.

Shared Wireless Access Protocol (SWAP): A protocol derived from the existing Digital Enhanced Cordless Telephone (DECT) standard. The SWAP specification describes wireless transmission devices and protocols for interconnecting computers, peripherals, and electronic appliances in a home wireless (HomeRF) network environment. The Shared Wireless Access Protocol referenced in the HomeRF standard should not be confused with the Wireless Applications Protocol (WAP), which is used primarily for wireless Internet access on cellular networks.

Shielded Twisted Pair (STP) cable: A type of network cable originally developed as an IBM cabling system. It was used to support Token-Ring LANs. The original specification for STP was released in 1984 and defined the 150-ohm STP cable types 1, 2, 6, 8, and 9 for support of frequencies up to 16 MHz. It also defined the 100-ohm Type 3 UTP cable, as well as Type 5 and 5J fiber-optic cables. Later, an enhanced IBM cabling system defined STP-A cable Types 1A, 2A, 6A, and 9A for support of a network standard called Fiber Distributed Data Interface (FDDI). The *A* suffix denotes the enhanced IBM cabling system. The original IBM cabling system was defined in IBM publication GA27-3773. The enhanced, or STP-A, cabling is defined in the ANSI/TIA/EIA 568A standard.

Simple Network Management Protocol (SNMP): A protocol used for network management and monitoring of network devices and their functions.

single-room system: A lighting control system designed specifically for the control of lights in one room, such as a home theater.

Small Office Home Office (SOHO): A generic term in the home networking industry that refers to products and services primarily intended for the small business community, and residential computer networks.

solenoid: An electromechanical device used with irrigation system control valves. A solenoid consists of an electromagnet that, when energized, moves a valve to the open position. A spring retainer moves the valve to the closed position when the voltage is removed from the solenoid.

spread spectrum: A digital radio transmission technology in which the bandwidth of the radio carrier signal is considerably greater than that normally required to convey the digital baseband information. Spread spectrum technology provides excellent resistance to interference from other radio signals, and intentional radio frequency jamming sources, at the expense of wider spectrum utilization.

spike: A short burst of energy superimposed on the AC power line, usually of very short duration. Magnitudes may range between 0.5 to 2000 volts, lasting for 8 to 10 milliseconds.

stand-alone: A term used to describe a home technology component's capability to operate independently of another component.

structured wiring: The term *structured cabling/wiring* refers to all the cabling and components installed in a logical, hierarchical way, with a central distribution box serving as a common termination point for all residential cabling. It's designed to be relatively independent of the computer or telecommunications network that uses it, so either can be updated with a minimum of rework to the home. Structured wiring uses a wiring/cable design in which each room in a residence is wired for dual coaxial and UTP cable outlets.

subwoofer: A type of speaker designed for use with surround sound multiple-speaker systems. These speakers are optimized to reproduce sound only in the lower frequencies of the audio spectrum region of 20 Hz to 80 Hz.

surround sound speaker system: A sound system that uses five speakers placed in special locations in the front and side listening areas, and a subwoofer to handle the low-frequency effects channel.

switcher: A video controller used with security systems to connect several surveillance cameras to a single monitor. The switcher can be programmed to cycle through all the cameras in a surveillance system, or to dwell on each camera for a specified length of time.

Telecommunications Industry Association (TIA): An organization accredited by ANSI to develop voluntary industry and home technology standards. The TIA's Standards and Technology Department is composed of five divisions, which sponsor more than 70 standards-setting groups.

telnet: A TCP/IP terminal emulation protocol that enables remote login (the capability to log in to another computer) from a local computer. Telnet is a common way to remotely control Web servers. Telnet is also a Unix command that starts the telnet program on a Unix computer. The telnet program is the software (based on the telnet protocol) that enables the user to log in to a remote computer, and use its resources. The most common reason for using a telnet program is to log in to a remote computer to use e-mail, or to run programs not available on the user's local computer. Remote login to a more powerful computer can provide access to a full range of Internet services and sophisticated programs that often cannot be run on a low-end personal computer. Telnet can also be used to access various library catalogs, and databases. Telnet programs (or clients) are available for a wide variety of computer systems.

three-way calling: A telephone subscriber service offered by local exchange carriers, long-distance carriers, and mobile wireless cellular carriers. It enables a subscriber to talk to two people, in two different locations, at the same time.

Three-way calling is initiated by calling the first party, pressing and releasing the switch hook to get a second dial tone, and dialing a second number. When the second party answers, the switch hook is pressed and released again. The three-way connection is then in place, and the call can proceed with all three parties connected.

three-way switch: A type of switch used to control a light fixture from two locations. An example of this type of installation is where a light needs to be controlled at the top and bottom of a stairway. A three-way switch is required at both locations.

tie wrap: A plastic tie for holding cables together or holding cable in place, usually in the interior of a wall, or in a cable tray.

Toslink: A popular consumer equipment fiber-optic interface, based on the Sony/Philips digital interface format (SPDIF) protocol, using an implementation first developed by Toshiba.

touchscreen: A type of technology that uses a cathode-ray tube (CRT), or a liquid crystal display (LCD) screen, as both a keypad and a display. Touchscreens are popular with kiosks, point-of-sale cash registers, library search terminals, and computer-aided design (CAD) and drafting applications.

Transmission Control Protocol/Internet Protocol (TCP/IP): The name applied to a suite of protocols developed by the Advanced Research Projects Agency (ARPA).

TRIAC (thyrister for AC applications): A type of solid-state device used in home lighting dimmer controls. TRIACs switch the AC voltage on and off, using a variable phase control principle, permitting a gradual increase or decrease in the percentage of the phase that is applied to the load.

true power: The actual power used by the load of an electrical circuit measured in watts. (Even though watts = volts x amps, apparent power is measured in VA to differentiate it from true power.) The relationship between true power and VA is represented by the cosine of the phase angle between the voltage and current. The relationship between real power (in watts) and apparent power (in VA) is expressed as power (watts) = cosine (phase) x apparent power (VA).

twisted pair: A communication medium consisting of two copper conductors insulated from one another, and twisted together to reduce the effects of electromagnetic interference. Twisted-pair cables comprise the majority of installed telephone and home LAN cables.

U

UF cable: A type of power cable designed for direct burial in the ground. It is often used for irrigation system valves and control devices. It has a typical operating voltage rating of 600 volts.

Uninterruptible Power Supply (UPS): The most widely used type of backup power system for servers, and other computers, in home networking. A UPS provides power to client computers and servers in a business location, or home office, for a brief period after a power failure to allow a gradual planned shutdown of computer systems. UPS specifications indicate the length of time as well as the amount of electrical power, the unit can deliver when commercial power is lost. The most important rating to be aware of is the volt-ampere (VA) rating, which is the product of the volts and amperes delivered to the load.

Universal Plug and Play (UPnP): A Microsoft standard. Using UPnP, a device can dynamically join a network, obtain an IP address, convey its capabilities, and learn about the presence and capabilities of other devices automatically. UPnP uses Internet and Web protocols to enable devices such as audio and video components, PCs, peripherals, UPnP appliances, and wireless devices to be plugged into a network and automatically know about each other. With UPnP, when a user plugs a device into the network, the device will configure itself.

Universal Serial Bus (USB): A computer standard for a serial interface designed to connect peripheral devices to personal computers. The USB Specification, Revision 2.0, covers three speeds: 480 Mbps, 12 Mbps, and 1.5 Mbps. The term *Hi-Speed USB* refers to just the 480 Mbps portion of the USB specification; the term *USB* refers to the 12 Mbps and 1.5 Mbps speeds.

Unshielded Twisted Pair (UTP) cable: A type of cable widely used in LANs. It has insulated conductors twisted together, leading to better electrical performance, lower crosstalk, and significantly higher bit rates than those available with untwisted cable. The grading system divides UTP cable into separate categories. The spacing of the twists in a given length, along with fire code and installation criteria, dictate the official category rating to which the cable belongs, and the bandwidth that can be supported.

V

Video Home System (VHS): A standard for video tape systems developed by JVC in 1976. With VHS processing, the intensity information (luminance) and color information (chrominance) are filtered separately to reduce the bandwidth, and are then remixed when played back. This filtering, and the fact that the playback format is composite video, serve to reduce the quality of the image.

Virtual Private Network (VPN): A secure, private communication connection that is used by two or more devices across a public network (like the Internet). These VPN devices can be either a computer running VPN software, or a special device like a VPN-enabled router.

Voice over IP (VoIP): A voice communications system that uses the Internet Protocol. VoIP systems send voice information in digital form as discrete packets, rather than in the traditional analog circuit-switched protocols of the public switched telephone network. VoIP is also a term used in IP telephony for a set of facilities managing the delivery of voice information using the Internet Protocol. A major advantage of VoIP, and Internet telephony, is that they avoid the tolls charged by ordinary telephone service.

W

Web pad: A wireless, portable, tablet-shaped, consumer-focused digital device usually combined with a touchscreen. It has a browser-based interface to simplify and enhance the home control experience. The two components of a Web pad are a portable LCD built into a tablet-shaped device, and a base station that is plugged in to the home network, which sends and receives wireless (RF) transmissions. Transmissions between the Web pad and the base station use RF, or wireless, home networking technologies, such as HomeRF, Bluetooth, or IEEE 802.11b, and consequently have a limited range of 150 feet.

Wide Area Network (WAN): A telecommunications network encompassing a large geographical area. Usually, WANs are considered to cover distances beyond a metropolitan area.

Wi-Fi: An abbreviation for Wireless Fidelity. Wireless networking equipment that meets the 802.11 standard is approved to use the Wi-Fi label. It is used generically when referring to any type of wireless 802.11 network device, whether 802.11b, 802.11a, dual-band, and so on. The term is promulgated by an organization called the Wi-Fi Alliance. Any products tested and approved as Wi-Fi Certified (a registered trademark) by the Wi-Fi Alliance are interoperable with each other, even if they are from different manufacturers. A user with a Wi-Fi Certified product can use any brand of access point with any other brand of client hardware that also is certified. Typically, however, any Wi-Fi product using the same radio frequency (for example, 2.4 GHz for 802.11b or 11g, and 5 GHz for 802.11a) will work with any other product, even if it's not Wi-Fi Certified.

World Wide Web Consortium (W3C): An organization created to promote interoperability and establish an open forum for discussion of the World Wide Web (WWW). In October 1994, Tim Berners-Lee, inventor of the Web, founded the World Wide Web Consortium (W3C) at the Massachusetts Institute of Technology Laboratory for Computer Science, in collaboration with the European Organisation for Nuclear Research (CERN), where the Web originated, with support from the Defense Research Projects Agency (DARPA) and the European Commission.

X

X-10: A powerline carrier (PLC) protocol that enables compatible devices throughout the home to communicate with each other via the existing 120 Vac wiring in the house. Using X-10, you can control lights and virtually any other electrical device from anywhere in the house, with no additional wiring. The protocol name is derived from *Experiment 10*, which was a successful communications protocol perfected by Pico Electronics of Scotland in the 1970s, and was originally intended to control audio turntables. The original X-10 patent expired in December 1997. X-10 is now an open standard, and many manufacturers are developing new and improved X-10 products.

XDSL: A term used generically to refer to any of the digital subscriber line variants, such as HDSL, SDSL, RADSL, ISDL, or VDSL.1.

Z

zone: A designated area established in an automated home design for programming lighting scenes, or audio source material from a remote source or control point.

zone bypass: A feature used with home security systems to temporarily disable interior sensor systems when the residents are present, but wish to secure the perimeter area.

Index

Ohm's Law, 36
Optoelectric, 38
Orthogonal Frequency-Division Multiplexing (OFDM),
   50

P

Passive matrix, 43
Personal Video Recorders (PVRs), 64
Phase Alternating Line (PAL), 47, 48
Pixels, 41
Plasma display, 43
Players, 16
Ported design, 32
Power, 36
Power rating, 35
Progressive scanning, 51, 59

Q

Quadrature amplitude modulation (QAM), 49

R

rear projection, 45
receiver, 23
receivers, 16
red, green, and blue (RGB), 68
Reflective, 39
RF carrier, 89
RF video distribution system, 89
RGB, 70
RGB sync on green, 71
RGBHV, 70
RGBS, 70
Ribbon speakers, 29

S

Satellite diplexers, 63
Satellite radio, 65
Satellite receivers, 23
Sealed cabinet, 32
Sequential Electronique Couleur Avec Memoire
   (SECAM), 47
Sheath, 38
Shielded single conductor, 38
Shielded twisted pair, 38

Signal, 21
Signal processors, 18
single-source device, 81
Single-zone, 81
Sound balance, 27
Sound cancellation, 26
Sound pressure level (SPL), 37
Sound reflection, 26
Speaker wire, 38
Speakers, 20, 40
Standard-Definition TV (SDTV), 50
Static contrast, 54
Stereo receivers, 23
Streaming audio and video, 65
Successive contrast, 54
Super Audio CD (SACD), 80
Surround sound, 77
S-Video, 69
Systeme Electronique Colour Avec Memoire
   (SECAM), 48

T

Target, 33
TFT (thin-film transistor), 43
Three-way speaker system, 34
Tone controls, 25
Toslink cable, 38
Toslink connectors, 72
Touchscreen controllers, 86
Transient contrast, 54
Transmissive, 40
Treble, 25
Tuner, 40
Two-way speaker system, 33

U

Unshielded twisted pair cable, 38
Unused channel, 90

V

Vestigial sideband (VSB), 48
Video Cable Terminations, 73
Video displays, 20
Video distribution pane, 87
Video distribution system, 87
Video projection, 44
Video scaler, 50